A History of Self-Harm in Britain

WITHDRAWN FROM
ST HELENS COLLEGE LIBRARY

Mental Health in Historical Perspective

Series editors:

Matthew Smith, Senior Lecturer, Director of Research (History) and Deputy Head of School of Humanities, University of Strathclyde, UK

Catharine Coleborne, Professor of History, School of Social Sciences, Faculty of Arts and Social Sciences, University of Waikato, New Zealand

Editorial Board:

Dr Allan Beveridge (Consultant Psychiatrist, NHS and University of Edinburgh, book reviews editor *History of Psychiatry*)
Dr Gayle Davis (University of Edinburgh, former book reviews editor of *History of Psychiatry*)
Dr Erika Dyck (University of Saskatchewan)
Dr Alison Haggett (University of Exeter)
Dr David Herzberg (University of Buffalo)
Professor Peregrin Horden (Royal Holloway)
Professor Mark Jackson (University of Exeter and Wellcome Trust)
Dr Vicky Long (Glasgow Caledonian University)
Professor Andreas-Holger Maehle (Durham University)
Professor Joanna Moncrieff (University College London)
Associate Professor Hans Pols (University of Sydney)
Professor John Stewart (Glasgow Caledonian University)
Professor Akihito Suzuki (Keio University)
Professor David Wright (McGill University)

Covering all historical periods and geographical contexts, this series explores how mental illness has been understood, experienced, diagnosed, treated and contested. It publishes works that engage actively with contemporary debates related to mental health and, as such, are of interest not only to historians, but also mental health professionals, service users, and policy makers. With its focus on mental health, rather than just psychiatry, the series endeavours to provide more patient-centred histories. Although this has long been an aim of health historians, it has not been realised, and this series aims to change that.

This series emphasises interdisciplinary approaches to the field of study, and encourages titles which stretch the boundaries of academic publishing in new ways.

A History of Self-Harm in Britain

A Genealogy of Cutting and Overdosing

Chris Millard
Wellcome Trust Medical Humanities Research Fellow,
Queen Mary, University of London, UK

palgrave
macmillan

Except where otherwise noted, this work is licensed under a Creative Commons Attribution 3.0 Unported License. To view a copy of this license, visit http://creativecommons.org/licenses/by/3.0/

© Chris Millard 2015

The author has asserted his right to be identified as the author of this work in accordance with the Copyright, Designs and Patents Act 1988.

Open access:

Except where otherwise noted, this work is licensed under a Creative Commons Attribution 3.0 Unported License. To view a copy of this license, visit http://creativecommons.org/licenses/by/3.0/

First published 2015 by
PALGRAVE MACMILLAN

Palgrave Macmillan in the UK is an imprint of Macmillan Publishers Limited, registered in England, company number 785998, of Houndmills, Basingstoke, Hampshire RG21 6XS.

Palgrave Macmillan in the US is a division of St Martin's Press LLC, 175 Fifth Avenue, New York, NY 10010.

Palgrave Macmillan is the global academic imprint of the above companies and has companies and representatives throughout the world.

Palgrave® and Macmillan® are registered trademarks in the United States, the United Kingdom, Europe and other countries.

DOI: 10.1057/9781137529626
E-PDF: ISBN 9781137529626
E-PUB: ISBN 9781137529633
Hardback: ISBN 9781137529619
Paperback: ISBN 9781137547736

This book is printed on paper suitable for recycling and made from fully managed and sustained forest sources. Logging, pulping and manufacturing processes are expected to conform to the environmental regulations of the country of origin.

A catalogue record for this book is available from the British Library.

A catalog record for this book is available from the Library of Congress.

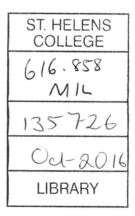

ST. HELENS
COLLEGE

616.858
MIL

135726

Oct-2016

LIBRARY

For Nathaniel and Lucas Rushton,
my extraordinary nephews, with love.

Contents

Acknowledgements

Thanks to Rhodri Hayward for the huge amount of intellectual and practical assistance over the past 6(ish) years. I could not have hoped for a better supervisor and then colleague. Thanks to Mark Jackson and Roger Cooter for their searching and constructive input throughout the process also.

Thanks to the Wellcome Trust for funding and beyond (with special thanks to Lauren Couch, Dan O'Connor, Chloe Sheppard and Nils Fietje, the people at the Trust from whom I have had most assistance, discussion and collaboration).

Thanks to Queen Mary, University of London, and especially the Centre for the History of the Emotions, directed by Thomas Dixon, for providing such a conducive environment for research, teaching and collaboration. Thanks to Miri Rubin, Head of the School of History at QMUL for her constant enthusiasm and encouragement.

Thanks to Katherine Angel, Elle Betts, Sarah Chaney, Thomas Dixon, Daniel de Groff, Bonnie Evans, Michael Gledhill, Åsa Jansson, Joel Morley, Rebecca O'Neal, Tom Quick, Lisa Renken, Emma Sutton, Jennifer Wallis and Tiffany Watt-Smith for reading parts of this, or other work, helping with mock funding interviews, for giving me many ideas, steering me away from excesses of theory and of detail, and generally helping me out.

Thanks also to Mark Jenner and Joanna de Groot, at the University of York, who first introduced me to new ways of thinking through history, sexuality and medicine.

To my parents for their continued and consistent support and interest. To my dear sister, Lizzie Rushton, and brother-in-law, Matthew Rushton, for everything. To my friends, especially: Ben Allcock, David Anderson, Jo Davey, Danni Haughan, Adam Hewitt, Tim Holmes, Adam Jamieson, Angus Macdonald, Rich Sunley, Will Turner and Joe Turrent, for their help, support, company, humour and loyalty over many years.

Thanks to my editor at Palgrave Macmillan, Jen McCall, and also to Jade Moulds, who together have steered me through this production process. Thanks to the two anonymous reviewers who have immeasurably improved the manuscript. Thanks also to Matthew Smith for insightful editorial comments and provoking me to think further about the present tense.

Finally, to Rebecca O'Neal, for support, ideas, discussion, inspiration and most of all, happiness: thank you.

This book was funded by Wellcome Trust Grant No. 089708/Z/09/Z. Many thanks to the Lothian Health Service Archives, Edinburgh, for permission to quote from their records.

Except where otherwise noted, this work is licensed under a Creative Commons Attribution 3.0 Unported License. To view a copy of this license, visit http://creativecommons.org/licenses/by/3.0/

Introduction: Self-Harm from Social Setting to Neurobiology

Self-harm is a significant mental health issue in the twenty-first century. The recorded rise in various behaviours, including deliberate self-cutting and self-burning, have been widely remarked upon and lamented.[1] Eminent cultural historian Sander Gilman has recently written of a global 'sharp public awareness of self-harm as a major mental health issue'.[2] The behaviour is usually said to be motivated by a desire to regulate feelings of intolerable tension, sadness or emotional numbness, and is almost always reported to be 'on the increase'; it is also often reported as a problem primarily affecting young women.[3] Despite a steady stream of books and articles on this emotive subject from the 1980s onwards – from psychiatrists, social workers and sociologists among others – there remains little meaningful historical analysis of this phenomenon.

This book sets out to provide such a history of self-harm in Britain in the twentieth century. It argues that to cast self-harm as an innate, eternal or transcendental practice (as much of the current literature does) is not helpful, historically speaking.[4] In fact it is decidedly ahistorical, as the core motivations underlying the practice of self-harm are seen as outside of history. This book shows how clinical ideas and medical diagnoses (such as 'self-harm') are intimately related to the specific, practical contexts in which they emerge and function. It also shows how shifts in concepts of self-harm correspond to much broader political trends. The central political shifts in this book are the ones that bring the welfare state into being after 1945, with nationalised industry and commitment to collective provision in housing and healthcare. This corresponds to an understanding of self-harm (overdosing) that is collective, communicative and socially embedded. The roll-back of the welfare state in the 1980s, coupled with the ascendancy of a more

1

individualised understanding of human beings as competitive and market-driven, corresponds to an understanding of self-harm (self-cutting) that is read as largely non-communicative and designed to regulate internal emotional states.

This book recovers and reconstructs, in detail, a clinical concern over an epidemic of overdosing presenting at British general hospitals between the early 1950s and late 1970s. This action is seen to be a response to, or communication with, a social circle or another person. This particular epidemic is part of a shifting chain of ideas about self-harming behaviour. These shifts partially come about through changes in the type and intensity of psychological and psychiatric attention focused upon self-inflicted injury (mostly overdosing) presenting at general hospitals. Self-cutting as a means of reducing internal tension emerges in very different circumstances – psychiatric hospitals dealing with inpatients – and is significantly influenced by North American psychoanalytic approaches. Once this archetype of self-harm is established, it begins to make sense as a model for the small minority of self-cutters (approximately 5–10 per cent) who present alongside the majority of overdoses at Accident and Emergency (A&E) departments in Britain. This book shows how dominant ideas about self-harm have gone through three broad phases during the twentieth century. From being seen in the early part of the century as a largely uncomplicated attempt to die, to a pathological communication with a social setting in the middle third of the century, to a method of regulating internal psychic tension that exists today. More recently, self-harm as tension reduction has begun to be understood in neurochemical terms, especially the notion of neurological triggering, as setting off an episode of self-cutting.[5]

The shift from understandings based upon social settings to ones based upon internal tension is of considerable political importance, given how it coincides with the collapse of consensus politics, the ascent of neoliberal economics, and the roll-back of the welfare state in the 1980s. It is a central contention of this book that the ways in which we make sense of our worlds, the categories and concepts that are available to understand human behaviour (such as self-cutting), resonate with and correspond to larger political constellations. The objects that seem so natural – that seem to have an independent, common-sense existence – are not outside of culture, politics, or ethics. In order to better understand this shift, this book reconstructs the middle phase of self-harm, alongside some stereotypes that preceded and succeeded it for comparison. Thus, the book aims to draw in detail an explanation of self-harm that relies upon the 'social setting'. This will establish a striking contrast with an explanation that has displaced 'the social' with explanations based on internal emotional states – which become increasingly expressed in neurological terms. The idea that somebody might damage themselves as a communicative act emerges in an influential way during the 1950s. Increased provision of psychological expertise at general hospitals makes available the explanation that people deemed to have harmed themselves might in fact be communicating in a psychologically disordered manner. The self-harm is predominantly achieved by 'overdosing' – taking medication in quantities considered excessive, but rarely lethal. Most studies of this phenomenon in the 1950s retain the term 'attempted suicide', whilst also emphasising that death is not the intended outcome for most patients. However, in the 1960s and later a large number of new terms are proposed by psychiatrists and doctors to try to deal with the confusing idea that 'attempted suicides' are not actually attempting suicide. Terms such as 'self-poisoning', 'parasuicide', 'pseudocide' and 'propetia' (rashness) are all put forward in order to deal with this confusion. However, the most common throughout the period remains 'attempted suicide'. The prominence of this supposedly communicative act increases in step with the level of psychiatric expertise available to general hospitals. This includes explicit efforts by the Ministry of Health to promote referral of attempted suicide patients to psychiatrists after suicide attempts are decriminalised in 1961.

This clinical and public-health concern begins to diminish in prominence from the late 1970s onwards. The generic category 'self-harm' comes increasingly to refer to self-cutting, seen not as a communication or appeal for help, but as a method of regulating internal tension or dispelling a sense of emotional deadness. It has recently been argued in a review of non-suicidal self-injury that whilst communicative and interpersonal models have been proposed, 'the affect regulation hypothesis has received the greatest amount of empirical support'.[6] Thus, the archetypal meaning of the label 'self-damage' or 'self-harm' shifts from self-poisoning as a communication, to self-cutting (and burning) as emotional control. Overdoses are now broadly conceived (outside of casualty-department-based epidemiological studies) as genuine attempts to end life.

To take just two examples of this displacement, the influential cultural psychiatrist Amando Favazza defines self-injury as: '*the deliberate, direct, alteration or destruction of healthy body tissue without an intent to die.* This construct excludes excessive dieting, pathological anorexia, acts committed with an intent to die, overdoses or ingesting objects and substances, body sculpting by drugs or weightlifting, risky behaviors, and cosmetic surgery (a topic for another book)'.[7] Thus overdoses (along with many other practices) are excluded. The fifth edition of the *Diagnostic and Statistical Manual of Mental Disorders* (*DSM-5*) includes a discrete self-harm category for the first time (rather than self-harm figuring only as a symptom of other disorders); 'Non-Suicidal Self-Injury' is described as 'intentional self-inflicted damage to the surface of his or her body', which again rules out overdosing.[8]

During the late 1990s there emerge several analyses of the ways in which the social setting has been displaced – including Nikolas Rose's provocative question around the 'death of the social'.[9] Roger Cooter has recently written that when 'humanness is flattened to the biological, the salience of the social disappears altogether'.[10] Sociologist Michael Halewood tentatively argues that scholarly discussions of '*the* social' only begin to appear in the early 1980s (for example Jean Baudrillard's work), when chronicling its supposed decline.[11] This fits in broadly with the chronology advanced here. The present book analyses one part of the idea of 'the social' as it is conceived and fabricated around an act of self-harm. I do not pretend to exhaust all possible concepts of the social, but such a narrow approach enables a clear idea of one particular and influential 'social setting', showing how it comes into renewed focus after 1945 and is then displaced from the 1980s onwards.

Analysis of the short heyday of overdosing as self-harming communication – between the early 1950s and late 1970s – can show how clinical objects are fundamentally tied up with the administrative practices and conceptual frameworks available at certain points in history. Important light is thus shed upon the relationship between the seemingly self-evident objects that populate our daily lives and larger shifts in the dominant explanatory frameworks in any given cultural system. This book is based primarily on the study of psychiatric research publications. The predominant focus is therefore on the ideas of psychiatrists and the clinical and administrative practices they describe. However, the broader political context and its resonance (especially around welfare and collective social responsibility) should not be forgotten and is flagged up where appropriate. The short career of this epidemic illustrates a far-reaching shift, from social and communicative understandings to ones based upon internal emotional tension, and then on to neurological ones. We must be clear about the ways of thinking that we are leaving (have left?) behind if we are to engage in an informed, ethically aware way with these changes. Self-harm presents an opportunity to track the ways in which certain influential understandings of behaviours are embedded in, and help to structure, their varied historical contexts. Such broad contextual and conceptual shifts have important political consequences – we must bring them into focus, undercutting their status as natural or 'common sense' – before we can engage with them politically, ethically and morally.

By linking attempted suicide and self-harm in this way, it might be argued that I am confusing or mixing up phenomena that should be kept distinct: attempted suicide should not be mixed with deliberate self-harm, or self-cutting should be understood as distinct from overdosing, to take two common objections. However we must not presume today's categories to be eternally valid. Instead of taking current categories at face value, this book analyses how the various assertions of difference and stereotypes come into play and how they are transformed over time.

The overarching aim of this book is to show one of the ways in which we have lost certain social, interpersonal perspectives in favour of individualised explanations based upon internal emotional states. It places professional, clinical analyses of this behaviour into detailed historical context, drawing upon the approaches of historical philosophers Michel Foucault and Ian Hacking. This 'genealogical' method seeks to analyse the rise of these behaviours and behaviour categories by connecting them to wider historical, intellectual and administrative contexts. It draws upon Hacking's insight about how people come to experience themselves through the concepts available to them at a particular point in time – what he calls 'making up people'.[12] This book charts the making up (and then part of the unmaking) of a certain type of attempted suicide, a cry for help, in a specific historical context. The idea that an informal arrangement attaching a psychiatric consultant to a casualty department, for example, could become an important part of a national public health problem seems counter-intuitive. However, it is this mix of small shifts (what Foucault terms 'micro-physics' or 'capillary power'), with an awareness of the overarching intellectual approaches of the time that shows how the ways in which we make sense of the world are continually shifting.

As well as its concern with reconstructing the social setting around self-harm in order to further emphasise its relative absence in the present, this book attempts three other interventions in the history of medicine. These are, broadly: to complicate the shift in psychiatric care from asylum to community; to analyse the role of the law in mental healthcare; and to explore the production of gender roles and sexism in mid-twentieth-century psychiatry. This book challenges current understandings of the history of psychiatry by interrogating the supposed move from 'asylum' to 'community care'. This shift is usually traced back to (then health minister) Enoch Powell's 'Water Tower' speech in 1961. Here, Powell casts the mental hospital ('isolated, majestic, imperious, brooded over by the gigantic water-tower and chimney combined') as a relic of the past. Instead, he claims, care for the mentally ill is better provided in 'the community'. (This model of care is much cheaper; as the asylums are phased out, no new money is earmarked for investment in community services.)

This binary of asylum-community underplays the mental healthcare provided at general hospitals. It is in these institutions that attempted suicides are treated for the physical damage but increasingly, as the century progresses, for the mental side of treatment, too. There are complicated interactions between mental and physical medicine inside general hospitals – through separate psychiatric wards, mixed wards, mental annexes, consultant and liaison psychiatrists and mental observation wards (something of a relic from the old poor law/workhouse hospitals). Through these varied institutions, mental medicine evolves in ways that are simply not captured by the tired binary of 'asylum-community'. Self-harm is perfectly placed to disrupt this simplistic but enduring attitude to the history of mental healthcare in the mid-to-late twentieth century.

This book not only revises understandings of the history of psychiatry in general, but also shows how legal changes and mental-health policy are absolutely crucial to the visibility and impact of these self-destructive behaviours. It shows how the Mental Health Act (1959) and the Suicide Act (1961) are linked. The former removes all legal restrictions for the treatment of mental illness in general hospitals; the latter decriminalises suicide and attempted suicide. Both are rooted in the same concern for appropriate psychological treatment without legal intervention as far as possible.

Finally, this account of self-harm analyses stereotypes of the actors supposed to perform the behaviours, with predominant focus on overdosing. Overdosing becomes highly gendered. The idea of an overdose as a cry for help draws upon ideas of feminine manipulation and emotional blackmail. The subsequent gendering in 'self-cutting as tension regulation' works slightly differently, feeding off the idea that men project their anger outwards, whilst women focus inwardly upon themselves (an outward/inward divide that has significant debts to conventional sex-role stereotypes). Deliberate self-harm has also been explained as a result of the stresses on women entering the workplace or, as it was put in a 2009 documentary, women trying to 'have it all' from the 1970s onwards. These antifeminist and often outright misogynist assertions have been critiqued elsewhere regarding self-cutting, but not for overdosing, hence my focus on the latter.[13]

A note on the present tense

This book is written entirely in the present tense. To write in this way is a tactic, with an objective, in the same way that writing history in the past tense is a tactic. These choices are tactical because they suppose – or at least imply – a particular relationship between 'history' and 'the past'. The past might be defined – relatively uncontroversially – as 'things that have happened before now'. If this is conceded, then history is not about the past. If we see 'the past' as the things that happened before today – indeed all things that have happened before today, before this moment – then what we are talking about is practically infinite, a senseless mass, a morass of impossible detail, of inhuman complexity. The past conceived in this way is an idea, but also a limit: it defines the present by continually pressing up against it and by swallowing up every possible human event, action or thought as soon as it has happened. We cannot speak about 'the past' as a whole entity – where could we start, let alone end? We can only talk about parts of the past. We can abstract from it, mobilise it, deploy it, use it. By making it partial, by editing, omitting, emphasising, glossing over, unpicking and ignoring the vast majority of the past, we can make it comprehensible, turn it into a story. This is the basis of history – making stories, making sense out of the past. This much is also uncontroversial, at least in academic history, since the early twentieth century. But even this obscures something and achieves something. For the past is not 'sitting there' waiting to be dug up, or analysed, or unearthed. It exists because we or others have put a marker down, because we are conscious of things having happened. Because we make a gesture of differentiating 'now' from 'then'. It is only after this differentiation that we can say the past is 'there'; it is only 'there' because we are 'here' in the present.

And this is the point of this lengthy discussion: we make 'the past' by acting in the present. We continually make and re-make the past. But more than that, we make the past by doing history – history begins with that differentiation between the past and present. Historians must clear the ground so that they can speak. They (we?) do this in the present, according to present concerns, with their (our) present tools, with their (our) present capacities, vocabularies and ideas. The past is an idea that is projected by history. The past is also the foundation stone that we all lay in order to recount our biographies, our very sense of self. But, depending upon where we are in our lives, these histories are different. The conceptual vocabularies we have emphasise different things; different things become visible and available. Child guidance, psychoanalysis or attachment theory impress upon our lives a very different sense to that given by evolutionary psychology or genetics. In the former, the events that loom large in history are those of our early upbringing; in the latter we focus on an entirely different order of time – a different history – to explain the roots of events. Our present choices, our present possibilities are pushed into a past tense that implies fixity, solidity or stability, when in the very next moment new events or new conceptual frames could overturn that whole edifice. The present tense is deployed to avoid this implication. It emphasises that the story being told is being told in the present, according to the present concerns, and under present constraints.

To tell these stories in the past tense risks the implication that they are fixed: that they are gone, done, dusted and immutable. The present tense is unsettling because we are so used to thinking of the past as 'over'. But history is never over; it is always about the present. Paul Connerton writes that we 'experience our present differently in accordance with the different pasts to which we are able to connect the present'. This is undoubtedly true. But the reverse is also true: the pasts that we are able to connect to the present depend upon the material, intellectual and social conditions of that present. History is thus made in, and governed by, the conditions of the present in which it is created. History is the present use of the past.

To claim that history is about the past and not the present is to make a mistake, to confuse a claim for authenticity with a statement of ontology. In less technical language: we should not take it on trust that history is (its ontology) what it claims to be. History gains much power and prestige by laying claim to the past, but again makes this claim according to present conditions. Thus history exists in the present through its claim to the past. So if the present tense is one tense (not necessarily *the* tense) proper to historical writing, as is argued here, so what? The answer to that is: some of the most important conditions of (present-based) history writing are political and ethical.

Writing history in the past tense carries the implication (or at least the possibility) that we are attempting to fix history in the past and to divorce it from the present. Given that the present is always saturated and thoroughly infused with political and ethical concerns, this amounts to an attempt to fix (make immovable) the political concerns of the present (as expressed through the history) by rooting them in the past. We attempt to give these concerns a sure, even immutable, foundation. The discomfort of writing history in the present tense is intended to keep permanently in view the politics going on here. I do not pretend that this history is fixed, or even that it is about the past. All the things described here happened, and are documented, in the conventional sense of having happened. I am not making this up. And yet, in another way, that is precisely what I am doing. I am making this history. I am performing it, researching it, selecting it. It is a product of my political, material, social concerns. All history is like this. It is an engagement with the present, under-girded by the materials available for thinking 'the past'.

There are probably many objections to this. I shall deal with the two most obvious ones here in a rather generic way: (1) it is confusing and alienating; (2) it is inconsistent and undercuts my argument. I shall deal with the first one by simply granting it. Writing history in the present tense is initially confusing and unsettling. It is sometimes labelled 'journalistic', which is revealing, but intended to mean 'unbefitting of historical scholarship'. It is a commonplace or cliché that 'journalism is the first draft of history' and, actually, it is precisely the provisional nature of a 'draft' that I would like to preserve. This is not to say that I believe this book to be slapdash, rushed or careless. It is to say instead that I want to be clear about this book's provisional nature, that it is a story from a certain place, at a certain time, with all the practical and intellectual constraints that this entails. The unsettling nature is also something that works towards my argumentative goal: I want to make people think about the distinctions that are concealed in the use of different tenses. Whenever the narration of a 1960s event in the present tense jars, I want to provoke a little reflection: 'this is happening now, not then' in the sense that this history is being understood, disagreed with, digested, made and remade according to twenty-first-century political concerns. It is not a simple reflection of events in the 1960s. I want to be clear as well as unsettling, and I hope I can be both. In any case, some feedback I have obtained says that the initially unsettling nature of the present tense does pass, and it becomes just another story told in the present – immediate and happening now as history does.

The second objection is perhaps more serious, and certainly more specific. I am dealing with one psychological category that has delimited shelf life: 'overdosing as a cry for help'; and another that remains very much with us in many societies: 'self-cutting as tension release'. It might be said that by talking of 'overdosing as a cry for help' in the present tense, I am implying that it is 'still here' and thus undercutting my argument about its specificity and context-dependent nature. But what does it mean to say that 'overdosing as a cry for help' is no longer with us? One could draw a comparison with nineteenth-century hysteria, characterised by catalepsies, palsies, fainting and paralysis. These behaviours, these conceptual understandings, are clearly no longer available as a widely understood pattern of behaviour. But this is to miss the point of the present tense: it is not saying that the objects described are here in the present, but that the description of them is occurring in the present. The historical descriptions of Bismarck's Germany, to take one example, are very different in 1950 to those in 1920. History is the understanding and abstracting and creating of 'a past' in the present. The present of 1950 has rather different historical concerns around Bismarck's Second Reich compared to those in 1920. My use of the present is not meant to imply that all events are happening now, but that our understanding of all events is happening now, and is always unfinished.

When I claim that history is happening now, it is also not meant in the trite sense that the past has material effects: for example, I fell over yesterday, and today my arm is broken. Instead I want to convey the sense that history – human sense-making, story-telling – is always properly thought of as happening in the present. When we forget that, we forget that human histories are always political, and always of the present. Writing in the present tense is not the only way to make this clear, but it is one way. It is not that the past (or reality) 'does not exist', or any other parodic nonsense often imputed to those ('postmodernists') who reflect critically upon the function and ethics of history. It is instead to argue that human understanding of the past is happening in the present. I want to write history with this awareness. I want to write history that is honest about its storytelling, its present function, and not confuse the mobilisation of the past with the implication of permanence and fixity. Perhaps in a different present I could write in a self-conscious, caveat-filled past tense. But today's present has the humanities under attack and the rolling back of the state's responsibilities. In addition – and most troublingly for historians – the present involves a nationalistic project under the guise of 'history as fixed facts' in schools. I cannot write in the past tense given what it implies in this present. Joan Scott argues forcefully and cogently of the value of history as critique, as work that can 'make visible the premises upon which the organising categories of our identities ... are based, and ... give them a history, so placing them in time and subject to review'.[14] The present tense makes this current, political project more obvious.

Here, I can be honest about my own political commitments, and I return to them throughout. This book details a history in which the social setting, and the agencies of state that nourish and buttress it, are present to an extent that is difficult to imagine today after the triumph of a market-driven and competitive understanding of human nature. What starts as the history of a psychiatric category runs parallel to the substantial disappearance of the idea of the social setting. The argument is that the concepts with which we populate and navigate our lives are related to political concerns. Human behaviours are vast and myriad, but they stabilise and congeal in certain ways, in certain objects, at certain times. The contrast between the present of the 1950s and that of the 1980s is here deployed in the 2010s to make clear that the retreat of the social setting has political significance.

The most entrenched and reactionary politics has ready recourse to the disguise that it is 'natural' or 'not political'. As feminists keep needing to reassert, 'the personal is political', and indeed the 'psychiatric' the 'medical' the 'historical' the 'social' – it is all political. The ways in which archetypes of 'self-harm' come to prominence and fade out might seem an unlikely place for a political statement. However, it is precisely where you do not think politics is happening that it needs to be exposed. This book is about how we arrived in this present with a particular set of ideas, stereotypes – cultural and intellectual shorthand with which we make sense of (a very small part of) our world. Again, if the familiar certainties and signposts of our lives (from self-harm to neoliberal human nature) are in fact made and remade by human action, then they are up for ethical debate.

Textbook emergence

Various forms of self-harm, including 'overdosing as communication' and 'self-cutting as internal tension regulation' are not eternal, ever-present, or rooted in an unbroken undercurrent of emotional response. Therefore, the specific emergence of these different stereotypes can be introduced through analysis of successive editions of three popular British psychiatric textbooks.

The *Textbook of Psychiatry*, written by David Kennedy Henderson and Robert Dick Gillespie, becomes known simply as 'Henderson and Gillespie' over ten editions and 42 years between 1927 and 1969.

Maxwell Jones remembers Henderson as 'the great high priest of psychiatry' at Edinburgh in the early twentieth century, while Gillespie is a brilliant but ultimately tragic figure who commits suicide in 1945.[15] Willy Mayer-Gross, Elliot Slater and Martin Roth's *Clinical Psychiatry* is also a standard textbook over three editions between 1954 and 1969. It is written, according to Slater, because '[t]he textbooks available at that time were either not very comprehensive or not all that good. The American ones were mainly full of Freud, or Adolf Meyer's psychobiology. Henderson and Gillespie was rather an old-stager'.[16] The emergent phenomenon of 'attempted suicide as cry for help' can be tracked, and its underpinnings glimpsed, through the editions of these texts, written as aids for trainee psychiatrists and general practitioners, as well as reference works for specialists. In both books, the epidemic of overdosing emerges as attempted suicide is being rethought and detached from completed suicide. It is transformed from a symptom of mental disorder to an object of scrutiny in its own right. Finally, Myre Sim's *Guide to Psychiatry* is useful for tracking the emergence of self-cutting for two reasons: not only is he well-informed in the field of self-harm (having published on the psychological aspects of poisoning), but his *Guide* runs to four editions, from 1963 to 1981.[17]

In Henderson and Gillespie, the principal references to suicide and suicidal behaviour in the first five editions (1927–40) concern the need for vigilance when caring for patients diagnosed with conditions such as 'depression' or 'involutional melancholia'. Suicide appears here as one possible outcome of psychiatric illness, a potential final symptom.[18] This is reproduced throughout the five editions up until 1940, with no effort to establish any differences of intent between people who succeed in their attempts and those who fail. The preface to the 1944 edition notes that 'remarkable progress...has occurred in psychiatry in recent years', which marks 'a new epoch in medicine and emphasises what psychiatry has for so long been doing – treating the individual in his social setting and making allowance for psychological as well as physical factors'.[19] In the chapter on 'Manic-Depressive Psychoses' there is this new material: 'We have been impressed by the large proportion of cases of attempted suicide admitted to the Royal Infirmary, Edinburgh, and Guy's Hospital, London, who have never previously seemed to require psychiatric guidance or control. The rapidity with which recovery occurs is also a factor to be noted and is in striking contrast to the prolonged treatment of the average case of depression'.[20] This emergent object is tentatively cast as a new (short-term) form of depression, appearing under wartime conditions, which emphasise the social setting's relevance to treatment.

This new object is distinctive, according to Henderson and Gillespie, because of the lack of previous psychiatric contact with the patients. Indeed, naming the hospital clinics serves to clarify that these attempted suicides are first assessed at general, rather than psychiatric hospitals. They are also 'struck by the trivial nature of the precipitating factors in some cases'. For example,

a husband requested by his wife to sleep for the time being in an attic to make room for a guest; a girl who had been 'walking-out' with a soldier of whom her father disapproved, so that being afraid to return home she walked into the Thames, near London Bridge, instead.

A supposed attempted suicide and bafflingly trivial interpersonal conflicts become visible at certain general hospitals. An element of communication is also noted in some cases, but this does not map neatly onto the division between those who survive and those who do not: 'Sometimes spite enters as a basis for the suicidal gesture, but it is a gesture which is sometimes carried to the point of successful self-destruction'.[21] These changes are linked on an intellectual level to the commitment to treat the individual in his or her social setting, which can potentially bring to light such conflicts. However, it is also down to the availability of informal psychiatric scrutiny in a clinic, outside of a mental hospital. After Gillespie's suicide, the sixth, seventh and eighth editions (1944, 1950, 1956) see radical changes in the authorship of the textbook. Henderson edits the 1950 version alone, and brings in Ivor R.C. Batchelor to assist with the 1956 edition. Despite these changes (and the fact that Batchelor publishes a number of articles on the subject between 1952 and 1955), the above text concerning 'attempted suicide' remains the same.

In the 1962 (ninth) edition, suicide and attempted suicide are clearly separated: 'Attempted suicide is much commoner than suicide in Western communities'. The idea that attempted suicide is separate from, more common, and less likely to be registered than completed suicide are key characteristics of the clinical object. Under the heading 'suicidal acts' attempted suicide is raised to the status of 'a social phenomenon of great importance and a concern not only to psychiatrists but to society as a whole'. They refer to Erwin Stengel's work on the social aspects of suicidal attempts, which leads to the suggestion that 'those who attempt and those who commit suicide constitute two different populations'. They note that Stengel's differentiation has an important gendered dimension: '[T]he majority of those who commit suicide are males while the majority of those who attempt it are females'. However, Henderson and Batchelor are not convinced that the populations are completely separate. They allow that the populations overlap, and that it would be unwise 'to draw any sharp distinctions between attempted suicide and suicide itself. ... No firm line can be drawn between suicidal gestures and suicidal attempts'. Nevertheless, they are broadly supportive of Stengel's project, arguing that 'emphasis on the appeal function of suicidal attempts and on the participation of the patient's group very properly draws attention to social aspects of individual suicidal acts'.[22]

The final (tenth) edition is published in 1969, edited by Batchelor alone after Henderson's death in 1965. Many studies of attempted suicide are mentioned; prominence is given to 'a notable increase in Britain of cases of self-poisoning, particularly with barbiturates and more recently with tranquilizing and other psychotropic drugs. ... The majority of these acts are impulsive: they are often the response to a quarrel or other frustration of a temperamentally unstable or psychopathic individual'. Batchelor quotes Neil Kessel (who works at the Royal Infirmary of Edinburgh in the early 1960s but overlaps with neither author):

> Kessel (1965) stated that 'for four fifths of (these) patients the concept of attempted suicide is wide of the mark ... what they were attempting was not suicide.' Certainly that there has been an attempt to seek attention and to manipulate the environment is often obvious: but Kessel goes too far in recommending that 'we should discard the specious concept of attempted suicide'.[23]

'Attempted suicide' has become a distinctive object within the field of suicidal behaviour, as a category that emphasises its potential as communicative with a social setting. Henderson and Batchelor are never quite convinced that it deserves a fully independent existence to the extent of Stengel or Kessel, but they certainly acknowledge its prominence post-1945.

The three editions of Mayer-Gross, Slater and Roth's *Clinical Psychiatry* show a similar pattern. In 1954 the authors note that '[s]uicide, or the attempt at it, is often the first alarming symptom of a depressive illness; it is the first and last symptom of many depressive illnesses'. They are clearly aware that there exists a less genuine class of 'attempts', affirming straight afterwards that '[i]n most cases [of depression] these attempts are desperately earnest'. The diagnoses most strongly associated with suicide (as a symptom) are depression, schizophrenia and psychoses in the aged.[24]

In the second edition (1960) there is a new section devoted to suicide where two separate objects are visible: '[A]ttempted suicide is estimated as occurring with a frequency of four to eight times that of consummated suicidal acts'. Its distinctiveness from completed suicide is again mapped onto gender: '[A]lmost without exception the rates for men are higher than women while the reverse holds for attempted suicide'. Again, Stengel is mentioned as having 'emphasised the "appeal" character of attempted suicide, the ambiguous or "Janus-faced" attitude directed at once to the reformation of human relationships and towards death'. Once more, the textbook authors are not wholly convinced that the objects are truly discrete, arguing that 'although it would be unwise to ignore the appeal element in a suicidal attempt, it would be more dangerous to over-estimate it'.[25]

In the third edition (revised by Slater and Roth after Mayer-Gross' death in 1961), the above material now merits its own subheading of 'Attempted Suicide' in a new chapter on 'Social Psychiatry'. Slater and Roth acknowledge '[t]he point made by Stengel and Cook (1958) that these are two separate but overlapping populations is now widely accepted'. They also refer to Kessel's argument that 'attempted suicide is not a diagnosis and not even a description of the behaviour of great numbers of cases coming for treatment under this heading even when the behaviour is clearly a deliberate act of self-injury and not accidental'. They mention three studies of incidence: Kessel's in Edinburgh, Stengel's in Sheffield and Farberow's and Shneidman's in Los Angeles.[26]

Thus a clinical object named 'attempted' suicide is articulated in two standard psychiatric textbooks after seemingly being brought into focus by Erwin Stengel and associates during the 1950s. Stengel does not create this object in any simple way; even without the 'trivial' precipitants in Henderson and Gillespie (1944) and implied 'non-earnest' attempts in Mayer-Gross, Slater and Roth (1954), it must be emphasised that these ideas do not spring from nothing, yet are also significantly novel. Crucially, it is not until after *Attempted Suicide: Its Social Significance and Effects* (1958), by Stengel, Cook and Kreeger, that the textbooks take a coherent position on this phenomenon where the communicative or appeal aspect is definitively acknowledged. Similarly, psychological clinicians from the 1960s onwards speak of an epidemic of suicidal behaviour by means of overdose that they believe to be novel ('currently fashionable') in important respects,[27] and in more recent psychological and sociological literature, the phenomenon of 'attempted suicide' is sometimes seen to begin to register around the 1960s.[28]

In both textbooks this recasting of 'attempted suicide' is based upon a shift from an action seen as a symptom or outcome of depression, to an object worthy of scrutiny more or less independently. This new object is first delineated simply by the arrival and survival of certain cases of injury presenting at general hospitals (predominantly after having taken an amount of medication). Through various interviews, investigations, follow-ups and assumptions, a social constellation is actively fabricated around the attempt, and meaning projected from the hospital into the social history of the patient. This awareness corresponds – as we have already argued – with a radical reimagining of the state's relationship with social welfare and its social responsibilities. The social setting and self-harm as social communication are brought to light (as we shall see) by particular groups – including social workers – who are part of that renewed commitment to welfare.

Overdosing becomes a serious public-health problem by the 1960s and, for a brief period, it is seemingly ubiquitous and constantly increasing at casualty departments around Britain. Then, rates begin to drop during the late 1970s, and by the early 1980s in Britain there is consistent acknowledgement of a particular sub-group of self-damaging patients. These people do not take overdoses but instead cut their wrists and forearms. It gradually becomes argued that they might be distinctive in more than just their choice of method. Increasingly, these patients are not seen as crying for help, but as regulating internal psychic tension by self-cutting. The social setting recedes, and the internal emotional life of the self-damaging patient comes into focus. This happens from the late 1970s onwards, again corresponding with a point at which the social responsibilities of the state are being rethought, with an emphasis on individual competition and self-support, rather than collective social support.

In the first edition of Myre Sim's *Guide to Psychiatry* (1963), there is a cursory mention of Stengel's and Cook's *Attempted Suicide* (1958), where he claims that 'the vast majority of those brought to the casualty department of a general hospital have what Stengel and Cook (1958) called an "appeal" character'.[29] In the second edition (1968) in a much bigger section on 'attempted suicide', Sim prefers Lennard-Jones's and Asher's 1959 term 'pseudocide' to Kessel's 'self-poisoning' because the latter 'by definition would exclude the considerable number who resort to self-wounding'.[30] Thus Sim is aware of a group of self-wounding patients, but he does not attribute to them any motivational or psychological differences. In the 1974 edition there are significant additions in the 'attempted suicide' section, including one headed 'Wrist-scratching as a Symptom of Anhedonia'. In it, Sim paraphrases the work of Stuart S. Asch, a psychoanalytic clinician from Mount Sinai Hospital in New York, whose study is based on psychiatric inpatients. Asch's work is mentioned by Barbara Brickman and the current author as part of the group of studies at the centre of this new profile around self-cutting. Michael Simpson's comprehensive literature review of this new kind of self-cutting in 1976 includes Asch's work as one of the 'classical studies'.[31]

Sim, paraphrasing Asch, mentions a profile of young women between 14 and 21 years of age who 'complain that they feel empty or dead inside and the striking characteristic is scratching or cutting their wrists'. He also mentions eating difficulties, promiscuity and abuse of drugs as common symptoms. Sim reports that Asch believes 'the cutting is a specific technique for dealing with both the rage and the depersonalisation'. The motivation and purpose of this behaviour as a means of affect-regulation (rather than communication) is clear. Sim obviously feels that the study has some value (or else why include it at all?), but he is sceptical about this profile, adding the comment:

> Wrist-cutting is common among males, particularly in prisoners and servicemen and there must therefore be a variety of interpretations. In the present writer's experience, girls who cut their wrists are generally from social classes IV or V [the two lowest], of limited intelligence and with a delinquent history, though a few do match those described above.[32]

In the fourth edition (1981), this passage about Asch's work is reproduced, along with another comment: 'Overdosing and self-injury [are] becoming an increasingly popular form of language'. He thus has both behaviours in there, and despite his section on Asch, he runs them together – likening them to hysteria: '[I]t was well-known that conversion hysteria became epidemic when doctors treated it as a legitimate disease'.[33] By the mid-1980s, self-cutting is an established – although contested – clinical object, and the significance of the communicative overdose is soon to diminish. Having sketched briefly the careers of self-poisoning and self-cutting, we can now look, in a more general way, at the issues involved in writing a history of self-harming behaviour.

Retrospective diagnosis and source-based confusion

The idea that individuals might harm themselves consciously, and not intend to die, seems timeless. However, this timelessness is often achieved by projecting our current ideas and concepts back into the past, making past ideas and events correspond to our current understandings. This practice, known as 'retrospective diagnosis' in the history of medicine, is problematic. Because it involves using terms in approximate, supposedly 'common-sense' ways, much of the following argument, setting up against such practices, may appear unnecessarily exacting or uncharitable to the scholars analysed. However precision in terminology and analysis of the assumptions underlying the choice of a particular term, are absolutely essential throughout this book. For it is only after much effort in defining the object of one's study, and reflecting upon the nature of the sources that one is using to talk about that object, that one can argue with confidence about the object's significance. How the object is defined governs the choice of sources to which one looks to provide evidence for it. To analyse self-harm is to enter a field littered with defunct and confusing terminology, as well as with vague attempts at 'catch-all' descriptions, so precision is not simply desirable but essential.

A discussion of suicidal behaviour at the London headquarters of the Royal Society of Medicine (RSM) in December 1988, is a good example of retrospective diagnosis. Norman Kreitman and Olive Anderson both present on the topic of 'suicide and parasuicide'. At this point, Kreitman is a distinguished psychiatrist, director of the Medical Research Council's Unit for Epidemiological Psychiatry in Edinburgh, and coming to the end of a successful, if unspectacular, career in psychiatric research. Olive Anderson is a fellow of the Royal Historical Society, and her seminal book, *Suicide in Victorian and Edwardian England*, has recently been published.

Kreitman's paper on prevention strategies strongly differentiates the terms 'suicide' and 'parasuicide', claiming that they 'differ in many radical respects' and that the differences between them 'outweigh their similarities'.[34] This is unsurprising: in 1969 Kreitman and three colleagues propose the term 'parasuicide' to describe 'an event in which the patient simulates or mimics suicide'.[35] As seen above, psychiatrist Erwin Stengel is credited by many with founding this new kind of concern around attempted suicide.[36] He sets himself up explicitly against the notion that 'a person who has attempted suicide...has bungled his suicide'.[37] Kreitman's terminological offering of parasuicide is one of many interventions reinforcing and rearticulating a distinction between acts aimed at causing death and those motivated by a more complicated and ambiguous intent. However, he is doing it in a specific context: his research from the late 1960s onwards focuses almost exclusively upon individuals conveyed to hospital after an overdose of medication.

Anderson's paper provides an historical gloss on suicide in Western Europe, from the late medieval period to Edwardian Britain. Perhaps prompted by Kreitman's presence, she includes a section on Victorians and parasuicide. This interdisciplinary attempt to communicate with clinicians on their terms – and at the RSM no less – is laudable. However the way in which she deploys the concept of parasuicide in an historical paper exposes the problematic relationship that sometimes obtains between history and psychiatry (and this despite her wider, careful and critical scepticism around labelling and suicidal behaviour). Her contribution here claims: 'Parasuicide is necessarily parasitic on a widely diffused assumption that self-harming behaviour should be responded to with help, sympathy and remorse, and this cultural breeding-ground flourished in Victorian England'.[38] It is important to be clear on what Anderson is doing here. She is making sense of the behaviour of people in Victorian and Edwardian Britain by using a term fashioned in a 1960s debate over communicative overdoses of medication.

Projecting parasuicide into the past in this way makes the behaviour (as defined by the 1960s terminology) seem timeless, ever-present and unchanging. The historical meaning of human action is flattened into current terminology, a description that is unavailable in Victorian Britain. In order for this analysis to work, the notion of a 'widely diffused assumption' stretches between the late 1980s and the Victorian era. In other words, the behaviour's meaning is cast as intended to procure 'help, sympathy and remorse' whether performed at the end of the nineteenth or the end of the twentieth century. This is a projection of the social setting – which is bound up with the core meaning of 'parasuicide' as well as a particular political period – into the past. The actions described by this term in one period are projected into the past. Assumptions and exclusions that create and isolate a stereotypical pattern of behaviour (its purpose, possible diagnoses and prognoses, the method employed, the gender, class or age profile, etc.) are transported from one context and imposed upon another.

Though Anderson seemingly makes parasuicide fit, the problems inherent in abstracting the term and projecting it into the past endure. She describes a nineteenth-century process in which the objective in assessing supposedly self-destructive behaviour is to 'distinguish the sham from the real', which is 'a daunting responsibility'.[39] This has superficial resonance with Kreitman's concerns, as when parasuicide is proposed it is claimed that 'what is required is a term for an event in which the patient simulates or mimics suicide'. However, parasuicide does not really speak to a debate between sham and real. The term differentiates between a largely uncomplicated wish to die and something much more complex than mere fakery: '[R]arely can his behaviour be construed in any simple sense as oriented primarily towards death ... this act, which is like suicide yet is something other than suicide'.[40] All this is lost in the redescription.[41]

The projection of current terms back into history leads to a second problem concerning historical sources. The set of historical phenomena (behaviours) understood through the parasuicide label are accessible because they are recorded and scrutinised in a particular Victorian context; these sources bear scant relation to those that underpin the 'parasuicide' term. This leads to a lack of awareness of the differences between the sources used to speak about suicidal behaviour – differences that have consequences for the historical objects described. One of the key sources supporting Anderson's claim that 'recorded suicide attempts far outnumbered registered suicides in Victorian London' is a one-off: 'Numerical Analysis of the Patients Treated in Guy's Hospital' (a general hospital) between 1854 and 1861. Information on 'attempted suicide' also comes from various police reports, as suicide is illegal in England and Wales until 1961 (see Chapter 3).[42] A term produced in the mid-twentieth-century around communicative overdoses brought to National Health Service (NHS) hospitals is unsuitable for understanding an attempted suicide composed of police records and a one-off hospital analysis. Combining information collected in different ways and for different reasons – and according to different definitions – to make a single object of concern termed parasuicide (under a different definition again) constitutes another problematic neglect of context.

Anderson is far from alone in making these leaps and is by no means the worst offender, but her interdisciplinary overstretch is a neat example that falls some way short of the thorough, nuanced work in her book. Projections like these make sense of a wide range of behaviours by rooting them in some eternal (and often unstated) emotional response or 'distress' or in a 'widely diffused assumption'. The history practised here aims to place understandings of behaviour in historical context, whether at the zenith of the welfare state or the ascendancy of neo-liberalism. It aims to show how practical arrangements and specific intellectual assumptions generate meaning in context. The past is to a great extent always a projection of present concerns, but this does not necessitate collapsing the past into present meanings.

These meanings are even more important when considering psychological categories that ascribe meaning to the actions of human beings who, themselves, are aware that they are being described and labelled in various ways. Through interaction with these powerful diagnostic labels, people can come to understand themselves through the motivations and emotions provided by the diagnoses. Telling someone that they are 'unconsciously crying for help' when they profess to be trying to kill themselves can change how individuals understand their own actions. Diagnoses can become much more than labels – they can form part of people's identities. If such descriptions are unavailable in a certain context, and the labels are different, the meanings produced are different.

After engaging briefly with some current accounts of self-harm, we can open up these conceptual and philosophical issues about descriptions of behaviour in the past, asking precisely what we mean by an epidemic of communicative overdosing before asking how it becomes available to historians in credible ways. The various sources of information (coroners' statistics, police reports, hospital records, etc.) that allow historians to access 'suicidal behaviour' are assessed, and the consequences that flow from using different kinds of information are outlined. These differences are a crucial part of the context. Since the nineteenth century, studies of suicide have been largely based upon well-established judicial registration procedures (coroners' statistics) from which a picture of 'suicide' is formed. No such registration practices exist for 'attempted suicide' in this period. From the late 1930s this phenomenon of overdosing emerges from hospitals, which are very different indeed from coroners' offices. The easy combination of material from very different sources (highlighted in Anderson's combination of hospital and police records to produce parasuicide) also occurs in the literature between coroners and hospitals, between suicide and attempted suicide. The distinction between two sources of information is erased.

We then turn to the specifics of our story. The context right at the core of this work is one which enables patients who arrive at hospitals after having suffered a physical injury to be assessed by psychological and psychiatric clinicians. It is this psychological expertise, and the assumptions contained within it, that enable the presenting physical injury (in this period, an overdose of medication) to be transformed into a pathological communication, a symptom of a disordered social environment, a simulation, or a cry for help. The possibilities for patients arriving at general hospitals to get consistent psychiatric assessment expand rapidly between 1950 and 1970. From the middle of the nineteenth century, much of British psychiatric practice is focussed upon the relatively remote mental asylums. The Mental Health Act 1959 is a familiar landmark in twentieth-century psychiatric history, representing a shift from this segregated model of provision. However, its impact in removing all legal obstacles for psychiatric treatment at general hospitals (where most attempted suicides are taken in the first instance, if at all) has been obscured by the dominant story of the failures of community care for the mentally ill coming out of psychiatric hospitals. In other words, the growing possibility of getting psychiatric treatment at a general hospital (which is not the community or an asylum) is absolutely crucial in the rise of this psychological object to national prominence.

The final section focuses upon the specifics of this psychiatric assessment. The place of the social environment and social relationships in mid-twentieth-century psychiatric thought (and especially psychiatric epidemiology) is of paramount importance in Britain. Thus, historically specific types of psychological expertise recast physical injury as a symptom of pathological social settings and relationships. Communicative self-harm emerges from a psychiatric tradition that focuses upon the social environment and psychiatric illness as communication. The idea of a cry for help might well have a broad intellectual ancestry, but it is structured and articulated by much more immediate intellectual and practical concerns.

Projections into the past: history and epidemic overdosing

Literature that engages historically or sociologically with the specific twentieth-century overdose epidemic is rare. The predominant approach presumes, explicitly or implicitly, an ahistorical constant which animates the so-called distress behaviour across time. Anderson's imposition of 'parasuicide' is especially clear in the second decade of the twenty-first century. The term never achieves sufficient popularity to be widely understood by non-medical audiences. As noted, the stereotype associated with it – the communicative overdose – has been largely forced from view by competing understandings of behaviour under the labels 'self-injury', 'self-harm' or 'self-mutilation'. As outlined above, from the 1980s onwards, the stereotype for intentionally self-harmful behaviour that is not directed at ending one's life involves young people cutting their forearms with sharp objects in order to regulate internal psychic tension. The overdose becomes recast as a genuine attempt to end life.

Digby Tantam and Nick Huband open their 2009 text with one of the clearest statements of differentiation between self-injury and self-poisoning, disqualifying themselves from commentary on the latter:

> This book focuses on people who repeatedly injure themselves by cutting, burning, or otherwise damaging their skin and its underlying tissue. This 'self-injury' is one of the two main types of self-harm, the other being self-poisoning with household or agricultural chemicals, or with medication. ... Self-injury and self-poisoning are often regarded as sufficiently similar to be considered as two facets of one problem. This fits with the observation that many of those who cut themselves also take overdoses, but it is not consistent with the very different cultural and psychological roots of self-injury and of self-poisoning.[43]

Making a related point, Jan Sutton, another twenty-first century expert on deliberate self-harm (DSH), uses questions-and-answers to analyse the term 'self-inflicted injuries'. This term is used generically in the media (and in Sutton's view, misleadingly) to talk about statistics from studies that include both cutting and overdosing: 'What sort of image does that [term] conjure up? Overdosing? I doubt it. Cutting? Highly probable'. She explicitly, and with confidence, closes 'self-injury' down into one specific behaviour: 'mention the word "self-harm" and it immediately conjures up images of people cutting themselves'.[44]

As well as this difference in the archetypal behaviour of self-harm, the dominant motivation from the 1980s onwards is seen to be the relief of internal tension.[45] In this way, as far back as 1978 Keith Hawton distinguishes between the motivations of self-cutters and self-poisoners (see Chapter 5).[46] Ideas of communication with a social circle or crying out for help become bound up in negative stereotypes about 'attention-seeking' behaviour, which is seen as unhelpful by many experts on self-harm. In Britain in 2004 and 2006 controversies erupt over self-harm where such negative stereotypes appear in national newspapers.[47]

In tune with Anderson's analysis, many experts argue that current self-injury concerns, the parasuicide epidemic of the 1960s and 1970s and Victorian attempted suicide are indeed largely the same thing and form an unbroken chain back into the past. Armando Favazza argues that self-mutilation has existed as long as humans have existed, finding it in 'Tibetan Tantric Meditation, North American Plains Indian mysticism and the iconography of Christ's Passion'.[48] Sutton claims that '[d]eliberate self-harm, parasuicide and attempted suicide [–] essentially they all refer to the same behaviour, and are sometimes used interchangeably'.[49] In the 1960s, eminent toxicologist Sir Derrick Melville Dunlop performs a similar projection using notions of hysteria:

> [D]ifferent generations tend in certain respects to vary in their patterns of behaviour. Thus, in Victorian times and in the earlier part of this century, in order to escape from a situation which had become intolerable, it was common, especially for younger women, to develop 'the vapours' – crude hysterias, fits, palsies, catalepsies and so forth. These hysterical manifestations are rare nowadays: it is easier to take a handful of tablets...not usually with any true suicidal motive but rather just to seek oblivion from, or to call attention to, unhappiness.[50]

Such a narrative involves a vision of the Victorian period different to Anderson's. However, the presumption of a pattern that only varies on the surface, if at all, is common to both methods of unifying the present and the past. They both use the past to anchor currently valid methods of sense-making.

That a relatively durable meaning might be stubbornly projected into many diverse behaviours – from catalepsies and fits to taking a handful of tablets – does not make it somehow eternal. That self-harm might 'seem to recur predictably' – to borrow from Joan Scott – does not insulate it from history, as not only are the 'specific meanings...conveyed through new combinations', but the very assumption of sameness needs to be investigated.[51] This kind of analysis equates current concepts (and their contextual baggage) to past phenomena produced in very different ways, for different purposes, through different practices, and understood through different assumptions. This conceptual 'presentism' cannot deal adequately with historical change. It must assume something real – that is, constant – underneath the different terms in different contexts.

How did self-harm become an object of study, and what kind of object is it?

Given these problems, how are we to proceed? Reversing Ruth Leys's formulation, borrowed for this section's heading, the first questions to be asked are: What precisely do we mean by saying that self-harm or parasuicide happened in the past? What kind of object is a communicative overdose in the past?[52] Having answered these questions, we can then discuss the implications of practising history in these ways. Finally, we can see how self-harm comes to be an object of study – that is, how it is recorded and treated as a statistical or clinical concern. This helps to explain how a specific epidemic of communicative overdosing becomes possible, prompting important questions about how and why human beings might behave in certain ways and at certain times.

In order to achieve a working definition of what self-harm is, it is useful briefly to revisit a debate initiated over a decade ago around chapter 17 of Ian Hacking's book, *Rewriting the Soul*, entitled 'An Indeterminacy in the Past'.[53] This debate focuses on whether a re-description of actions in the past using present categories (such as Anderson's use of parasuicide to describe actions in Victorian Britain) is legitimate, and whether it changes the actions. Hacking's questioning examples include: Are Canadian soldiers shot for desertion during the First World War, now sufferers from post-traumatic stress disorder (PTSD)? Is an eighteenth-century, 48-year-old Scottish explorer a child molester (then or now) for marrying a 14-year-old girl?

The legitimacy and consequences of various re-descriptions are analysed through Hacking's engagement with influential Wittgensteinian philosopher G.E.M. Anscombe's *Intention* (1957). In Anscombe's argument, the most relevant point to Hacking's project (and what emerges in the debate) is the focus upon context. Hacking engages with Anscombe's key example of a man pumping water. He states that

> [o]ne of the ways in which human action falls under descriptions is in terms of the way the action fits into a larger scene. The man's hand on the pump is going up and down. Enlarge the scene. He is pumping water. Enlarge the scene. He is poisoning men in the villa. As Anscombe makes so plain, the intentionality of an action is not a private mental event added on to what is done, but is the doing in context.[54]

Kevin McMillan's contribution to the debate makes this contextual point especially clear. He argues that we can get a handle on what social phenomena might be (for example, an epidemic of attempted suicide) by 'identifying and distinguishing them in terms of their historical, cultural or domain specificity'. He appreciates that this has consequences: 'An emphasis on specificity may make us chary of indiscriminate retroactive re-description. When applied, re-descriptions – particularly in terms of modern moral concepts – drag a complex and extensive practical, moral, epistemic and conceptual baggage in tow'.[55] It is an appreciation of this baggage that is crucial to understanding self-cutting, overdosing, and so on, in a fully historical way – to be wary of describing past actions with current concepts, or of collapsing them into one another.

Following this line of analysis means that socially directed attempted suicide cannot exist as a concept or pattern of behaviour independent of the institutional channels and professional scrutiny through which it is constituted. Specifically, this involves the increasingly consistent provision of psychiatric scrutiny available to patients presenting at general hospitals (as we shall see below). To separate the communicative self-harm from these practices would be to divorce it from its context. Following Allan Young's argument around the category of post-traumatic stress disorder (PTSD), it is argued here that self-harm 'is not timeless, nor does it possess an intrinsic unity. Rather, it is glued together by the practices, technologies and narratives with which it is diagnosed, studied, treated, and represented and by the various interests, institutions and moral arguments that mobilized these efforts and resources'.[56] Particular practices and technologies (new arrangements for psychological scrutiny) create attempted suicide as cry for help. As the contexts change, the objects change. This is not to say that people in the past have not used the terms 'cry for help' or 'attempted suicide' or that they were wrong to do so. However, they are not talking about, recording or accessing the same thing.

It is negligent to collapse this diverse richness into the psychological (or neurobiological or sociological) categories that happen to be current today. Adrian Wilson argues that 'concepts-of-disease, like all concepts, are human and social products which have changed and developed historically, and which thus form the proper business of the historian'. He describes the consequences of retroactive re-description, an approach

in which diseases throughout history have been identified with their modern names-and-concepts...the effect of this approach is to construct a conceptual space in which the historicity of all disease-concepts, whether past or present, has been obliterated. Past concepts of disease have simply been written out of existence; and the historicity of modern disease-concepts (or what are taken to be modern ones) is effaced, because those concepts have been assigned a transhistorical validity.[57]

This effort to homogenise and assimilate might well have a present utility, as well as broadly progressive political effects, as in the case of Canadian deserters and PTSD. However, in order for objects to have such a transhistorical and decontextualised existence, their conditions of production must be obscured; in other words, they only make sense outside of context – utterly unhistorical. Thus the meaning of an epidemic of communicative attempted suicide is more precisely stated: a specific understanding of behaviour, inseparable from its context. Having established the importance of context in general, the specifics can now be tackled.

This object emerges from of a complication of behaviour previously thought to be suicidal. Behaviour that presents at hospital ostensibly as an attempt to inflict death upon oneself is recast as a communication, as an appeal to a social setting. To understand how this attempted suicide comes to be a clinical, statistical and an epidemiological object, we must compare the historical and institutional contexts through which self-harm and completed suicide are accessed and analysed in Britain in the nineteenth and twentieth centuries. Mid-twentieth-century suicide strongly resembles its nineteenth-century counterpart as it is accessed through coroners' court records. However, objects called attempted suicide in the nineteenth and twentieth centuries (including communicative overdoses in the latter period) are separated by a profound difference, with far-reaching consequences.

Attempted suicide in the nineteenth century: asylums, and others

Information about attempted suicide in the nineteenth century comes from a variety of sources. Anne Shepherd and David Wright use the most popular set of source materials used to access these 'suicide attempts', county asylum records. They argue that these provide 'a useful comparison to the more frequently used coroners' reports that underpin most research on Victorian suicide'. They describe

a dominant and influential tradition of researching the history of self-murder from death certificates, coroners' reports, and official parliamentary statistics. We thus know a great deal about those who were 'successful', but much less about those who had 'failed' to take their own lives. Attempted self-murder remains relatively uncharted territory.[58]

Shepherd and Wright do not elaborate upon the differences between the various registration practices, nor on the consequences flowing from the different kinds of access to 'failed' or 'successful' objects in the past. However, they do suggest that the label 'suicidal' includes both 'real' and 'fake' attempts, and is therefore ambiguous.[59]

Åsa Jansson's conceptually precise study of suicidal propensities in nineteenth-century psychiatric literature and asylum casebooks demonstrates the fundamental relationship between recording practices and conceptual possibilities, concluding that there is no easy relationship between the adjective suicidal and the noun suicide in this period. This position is based upon a clear and consistent appreciation of the different recording and statistical practices that underpin them. The former ('suicidal') is based upon asylum statistics, the latter ('suicide') on coroner's statistics. These are collected in different ways, under different definitions, for different purposes.[60] Sarah Chaney's study of suicide at Bethlem (1845–75) is thorough and detailed, including sustained efforts to access and analyse meanings around attempted suicide.[61] Both show that nineteenth-century (and before) information about attempted suicide does not come from so organised and systematic a source as coroners, who record and categorise the dead, not the living attempter.

Twentieth century: observation wards and general hospitals

Erwin Stengel, influential commentator on communicative self-harm, does not use asylum statistics for his studies in the 1950s and 1960s. He begins his most influential researches through clinical work in mental-observation wards attached to general hospitals, places significantly associated with attempted suicide patients. These are parts of general hospitals where psychiatric scrutiny is available. This crossover point between general and psychiatric medicine, along with the inclusion of mental health services in the NHS, forms an absolutely crucial historical context for the emergence of this epidemic of overdosing. This object, named (somewhat misleadingly) 'attempted suicide' by Stengel, begins to grow. It is based upon the psychiatric scrutiny and assessment of patients brought to general hospitals after having suffered an injury presumed as self-inflicted (the majority of which are overdoses), principally at these mental-observation wards. It is increasingly recast as a pathological communication with a social circle or significant other. A number of psychiatrists, including Frederick Hopkins in Liverpool (1937–43), Stengel in London (1952–8) and Ivor Batchelor in Edinburgh

(1953–5) begin to exploit the uneasy cohabitation of general medical and psychiatric expertise in these 'secure' areas connected to various general hospitals.

Suicide statistics from coroners' court proceedings are thus fundamentally different to psychological analyses of attempted suicide from mental-observation wards. Despite this difference, it is still possible to connect suicide and attempted suicide in an abstract sense, at a level of competence: individuals wishing to kill themselves might survive by accident, or someone attempting to cry for help might die after causing an injury more serious than intended. Nevertheless, the data through which these objects are constituted – through which historians are able to study them – are not the same.

Hopkins and Stengel are aware of some differences. Hopkins, whose study forms part of Chapter 1, mentions in 1937 that there is 'no one authority to whom all [attempted suicide] cases must be notified'.[62] Stengel, combining observation ward records with extensive interviewing, laments in 1959 that 'there is no machinery for their registration'.[63] They both implicitly contrast observation wards with coroners' courts, and the laws that require suicide deaths, but not attempts, to be registered. However, this contrast primarily establishes the inadequacy of the former, rather than their fundamental difference. Later, during the 1970s, the context for self-harm shifts again: from general hospital A&E departments to psychiatric inpatient facilities. Again, the data available at these institutions are significantly different.

The work of Michel Foucault provides strategies for analysing the changing, historically specific technologies that produce 'objective facts' about the world. He claims that through analysis of these technologies, these practices, it 'can be seen both what was constituted as real for those who sought to think it and manage it and the way in which the latter constituted themselves as subjects capable of knowing, analyzing, and ultimately altering reality'.[64] The present book undertakes close analysis of specific practices and contexts to show how ideas of self-harm could function, for a time, in certain places, as an idea, a diagnosis, an epidemic. It also shows how ideas might change and correspond to broader political shifts around welfare provision, social work and the later rise of neo-liberal individualism. There is no attempt here to find the 'real' meaning or some unchanging emotional response that is expressed through varying cultures, but to appreciate the fundamentally historical character of concepts.

Having argued for the centrality of context, it is important to sketch out two specific contexts being drawn in increasing detail throughout.

General medicine and psychiatric expertise are persistently separate throughout this period, but the ways in which these approaches are separated undergoes radical change. The second context concerns social psychiatry. This particular conception of mental disorder, and the importance of social relationships in the aetiology of psychic disturbances, are vital parts of the credibility of an epidemic of communicative overdosing. A huge amount of intellectual and practical labour is invested in accessing the 'social settings' of people brought under psychiatric scrutiny. This is the same social setting with which the 'attempted suicide' is said to be communicating, and part of the social that falls away in the 1980s when 'self-cutting as tension reduction' begins to displace 'overdose as cry for help' as the archetype for self-damaging behaviour.

Separated therapeutic approaches

According to standard narratives mental medicine is largely separate from other branches through the geographically remote lunatic asylum from mid-nineteenth century Britain.[65] This insulation of psychiatric from general medicine is a key area in which change is sought during the twentieth century. A divide endures: the *National Service Framework for Mental Health* (1999) recommends that mental healthcare be provided by 'single-speciality mental health trusts' in urbanised areas, proposing a sharp administrative division between psychological and general medicine. Two liaison (general hospital) psychiatrists argue in 2003 that 'these mental health trusts threaten to repeat the mistakes of their 19th-century predecessors' by perpetuating the stigma of mental illness, and undermining the view that 'the distinction between physical and mental illness is conceptually flawed'.[66] Regardless, single-speciality trusts are again championed in a 2007 Department of Health Annual Report.[67] The mid-twentieth-century history of this divide runs through three acts of Parliament: The Mental Treatment Act 1930 allows non-certified treatment in county asylums; the establishment of the NHS (1948) brings mental and general medicine under the same administrative structure; the Mental Health Act 1959 removes all legal barriers to the treatment of mental illness in general hospitals. These developments are written into a smooth narrative of progressive integration, with 1959 as the culmination of the process.[68]

This simplistic narrative flattens the three decades between 1930 and 1960 into a smooth road away from legal constraint and the stigma of separation, and from asylums themselves in a process known as deinstitutionalisation.[69] Efforts to integrate the separated therapeutics of mental and general medicine form a crucial backdrop throughout this book, but instead of being smooth or teleological, this process is uneven, faltering and local. This separation of mental from general medicine is not inevitable, or rooted in some deep-seated consistent organising principle. It is the result of a number of complicated historical developments, and is sustained by specific practical and institutional arrangements.

Psychological scrutiny becomes entrenched in general hospitals, and the crossover between physical and psychological medicine forms the core of attempted suicide throughout the period. The shifting and specific arrangements that effect this crossover are described sequentially. It is worth reiterating here that these divisions are not self-evident, natural or inevitable; this will become clear as each arrangement is discussed in turn. This argumentative focus cuts across the standard asylum-to-community narratives in the history of twentieth-century psychiatry. Too close a focus on the well-tilled ground of 1959 obscures the significance of developments in general hospital mental-observation wards that significantly foreshadow the late 1950s attempts to combine psychological and general medical expertise.

During the early 1960s, studies emerge from various places (including London, Edinburgh, Birmingham and Sheffield) establishing attempted suicide as an epidemic phenomenon. This is principally because the opportunities for psychological scrutiny of patients presenting at hospitals with 'physical injuries' is increased by the changes and trends made explicit and further enabled in the Mental Health Act. Attempted suicide becomes an object of study through a transformation of physical injury into a psychosocial disturbance. That is, the injury that provokes admission to hospital is subordinated to a pathological social situation or psychological state. Patients arrive at hospital casualty departments, the most non-specialised part of the hospital system, due to a physical injury. After this has been assessed for its urgency, the patient might be treated with stitches or stomach-washing within the department, or transferred for resuscitation or surgery. It is only after this physical injury has been dealt with that the patient is investigated from a social-psychiatric point of view, and this is increasingly carried out by different medical professionals. Patients must consistently be referred for psychological scrutiny if the supposed cry for help is to emerge on any significant scale. This transformation thus rests upon two innovations: consistently applied arrangements focussing psychological scrutiny upon patients presenting with a physical injury at a general hospital, and the resources for intense scrutiny and social follow-up, to fabricate a credible social setting to which the attempted suicide is supposed to be appealing. (The strong differentiation between physical and psychological used to clarify this transformation might be unclear, unimportant or ambivalent for the patients, or anyone else who helps effect their removal to a hospital.)

Relating a physical injury to a social, domestic, romantic or familial context is time-consuming and labour-intensive, requiring interviews, questionnaires, social workers, follow-up and home visiting. The injury is not just contextualised, it is fundamentally recast as a symptom of this social setting. A specifically domestic social context is constructed in various credible ways by a newly influential profession of psychiatric social workers (PSWs). It is through consistent psychological scrutiny around general hospitals that suicidal intent is made complex and ambiguous, in a consistent and stable way. The possibility for this epidemic is, therefore, fundamentally contextual and historical. It is constituted and sustained by various possibilities for different kinds of scrutiny within a specific healthcare system. Changes in hospital organisation, mental healthcare provision, medical research and the law are all implicated in the emergence of this new object.

Stress, social psychiatry and psychiatric epidemiology

Just as the administrative separation of (and referral between) general and psychiatric medicine is important in the constitution of communicative attempted suicide, the type of psychiatric scrutiny focused upon the cases so referred is also important. This psychological object emerges through psychiatric epidemiology. This branch of psychiatry associates mental disorder with certain features of the environment – in this case, the social environment. There is a thriving field analysing the history of this psychiatric sub-specialism.[70] It is significant (and unsurprising) that a branch of mental medicine so concerned with social spaces and relationships interprets self-inflicted injuries as communications with that social environment. (It is important not to confuse the specialised, environment-focused usage of 'epidemiology' with the common usage of 'epidemic', meaning simply a high number of people affected.)

Ideas of stress and coping are integral to social psychiatry and psychiatric epidemiology in Britain. Mental disorder is embedded in social relationships and situations through notions of 'stress'. Mark Jackson's survey of twentieth-century stress research notes 'the capacity for the language of stress to clearly articulate the relationship between organisms and their environment...in debates about the social and cultural determinants of mental illness'.[71] The history of psychology traces 'stress' back through the work of Hans Selye (1907–82), whose General Adaptation Syndrome is based upon endocrinological experiments with mice; and Walter Canon (1871–1945), whose first famous experiments are with dogs (he later coins the phrase 'fight or flight' to describe responses to stress and establishes the concept of 'homeostasis').[72] This leads back to Adolf Meyer and his use of a 'life chart' to explain psychological disorder. Jackson cites the influential works of Harold Wolff, Daniel Funkenstein, Roy Grinker and John Spiegel as evidence that it is this psychosocial approach 'rather than Selye's experimental physiology that came to dominate clinical and epidemiological accounts of stress'.[73]

Stress gains prominence during the late 1960s and 1970s through psychological rating scales, especially the US-based work of personality theorist Raymond B. Cattell, and Thomas Holmes and Richard Rahe's Social Readjustment Scale and Schedule of Recent Experience (1967). In Britain, anthropologist George Brown and social psychiatrist Tirril Harris construct the Bedford College Life Events and Difficulties Scale in the 1970s.[74] Perhaps the most influential twentieth-century articulation of stress is found in Post-Traumatic Stress Disorder (PTSD), the genesis of which Allan Young meticulously charts through Veterans' Administration hospitals in the aftermath of the American war in Vietnam.[75] Rhodri Hayward argues that it is 'now a commonplace among psychiatrists, sociologists and historians to bemoan the ill-defined nature of stress and the theoretical fecundity that this sustains'.[76] Precisely this fecundity is the focus here, for stress is much broader than this particular historical thread. It is a key intellectual plank for the projects of social psychiatry and psychiatric epidemiology, through the links it makes possible between environment and mental disorder.

The necessity for a new model to guarantee psychiatric epidemiology is clear in light of traditional epidemiological concerns. Up until the Second World War, this approach makes most sense in the quest to control and prevent infectious diseases such as typhoid, cholera and influenza. After 1945, epidemiological methods are increasingly applied in psychology. Mark Parascandola argues that 'by the 1950s epidemiologic methods and thinking had expanded beyond the mere study of epidemics to human experiments testing preventative interventions, case-control observations in hospital patients, and the long-term study of generally healthy cohorts'.[77] The 'epidemiology of mental disorders' begins to make sense – as the distribution of mental problems within a defined area. However, this shift is controversial for some. In 1952, a professor of Bacteriology writes in 1952 of his fury at an undoubted debauchery of a precise and essential word, 'epidemiology' which is being inflated by writers on social medicine and similar subjects to include the study of the frequency or incidence of diseases whether epidemic or not[;]...an epidemic is disease prevalent among a people or a community at a special time, and produced by some special causes not generally present in the affected locality. Therefore, to speak of the epidemiology of coronary thrombosis, or of hare lip, or diabetes, or of any non-epidemic disease, is a debasement of the currency of thought. It is of no use saying that the word is being used in its wider sense. It has no wider sense.[78]

Michael Shepherd – the first ever professor of Epidemiological Psychiatry – points out that this is not a new concern. He cites J.C.F. Hecker's *The Epidemics of the Middle Ages* (1859) which, in addition to surveying the Black Death and the Sweating Sickness, also deals with an epidemic of 'disordered behaviour, the Dancing Mania [and] makes no distinction between epidemics of infectious disease and those of morbid behaviour'.[79] However, psychiatrist and anthropologist G.M. Carstairs, head of a research unit on the 'Epidemiology of Mental Disorders' is uneasy about the meaning of the word in 1959, noting that 'I find that this term "Epidemiology" is in the process of acquiring a new, specialised meaning which is at a variance with its generally accepted one: the study of epidemics. As a result I find that even with medical men the term "epidemiology of mental disorders" usually requires some explanation'.[80]

Carstairs glosses the history of psychiatric epidemiology in his application to head this research unit. Two key events are the 1949 annual conference of the Milbank Memorial Fund in New York and a review by Eric Strömgren from 1950. Carstairs also mentions a London-based 'international working party on research method in psychiatric epidemiology' in September 1958, which met to 'discuss, amend, and finally endorse a "canon" of research methodology', which is later published under the auspices of the World Health Organization.[81] Despite this, there remain serious conceptual issues with psychiatric epidemiology – namely the lack of a single agreed model to relate mental disorder to groups of human beings, rather than individuals.

Psychiatric epidemiology and social psychiatry begin to make sense in the twentieth century thanks to a broad and eclectic set of explanations under the terms 'stress' and 'distress', which are neither normal nor pathological. In the twentieth century, 'the social' is rearticulated through 'stress', 'distress' and 'coping' in new and pervasive ways as a source and broad canvas for psychological problems, so that by the early 1950s 'the psychiatrist ... is incessantly forced to consider the social relations of his patient'.[82] David Armstrong's *The Political Anatomy of the Body* (1983) contains perhaps the most compelling and wide-ranging demonstration of this in a British context. Armstrong's argument is structured by a shift from 'panoptic' to 'dispensary' medicine:

> the panoptic vision created individual bodies by objectifying them through their analysis and description[.] ... The new body is not a disciplined object constituted by a medical gaze which traverses it, but a body fabricated by a gaze which surrounds it[,] ... a body constituted by its social relationships and relative mental functioning.[83]

Further, the link between social relationships and stress is made clear by Armstrong through links with sociology: 'In psychiatry, sociology has provided a rich and diverse contribution to the extension of the medical gaze[;] ... theoretically it, together with psychology, has helped to define basic concepts, such as stress and coping. ... In short, sociology has reinforced the shift of the psychiatric gaze'.[84] In 1965 Neil Kessel expresses 'self poisoning' in the language of limits and 'coping': 'Nobody takes poison, a little or a lot, to live or to die, unless at that moment he is distressed beyond what he can bear'.[85] The idea that communication is central to mental illness is broadly characteristic of psychiatric thought after the Second World War. It is no coincidence that it emerges in the context of the state's efforts to manage the social setting, through social work and socialised medicine. In a context wherein collective responsibility for health and social security is established, this idea of health and disease as socially embedded and communicative is widespread. In fact, this idea becomes central to so-called 'anti-psychiatry' as much as mainstream psychiatric thought.[86]

In Jurgen Ruesch and Gregory Bateson's *Communication: The Social Matrix of Psychiatry* (1951), Ruesch touches upon the practical shifts mentioned above, noting that '[p]sychiatrists have moved out of the enclosing walls of mental institutions and have found a new field of activity in the general hospitals of the community and in private practice'. Importantly, this leads to the argument that 'it is necessary to see the individual in the context of his social situation'. In fact he goes even further, claiming that it is 'the task of psychiatry to help those who have failed to experience successful communication' and that '[p]sychopathology is defined in terms of disturbances of communication'. Ruesch admits that such a formulation might be a little surprising, but that the sceptical reader need only open a textbook of psychiatry to find that terms such as 'illusions', 'delusions', 'dissociation' or 'withdrawal' in fact 'refer specifically to disturbances of communication'.[87]

A decade later, Thomas Szasz's *The Myth of Mental Illness* (1961) casts hysteria as an archetype for psychiatric practice, an 'historical paradigm of the sorts of phenomena to which the term "mental illness" refers'. In other words, hysteria is not only an excellent example, but the definitive example. One of the pivotal chapters in this foundational text of anti-psychiatry is 'Hysteria as Communication'. Similar to Olive Anderson's comments about distinguishing sham from real in Victorian attempted suicide, Szasz argues that hysteria 'presents the physician with the task to distinguishing the "real" or genuine from the "unreal" or false'.[88] This also links up to Derrick Dunlop's (1967) and Raymond Jack's (1992) associations of self-poisoning with hysteria. Ideas around communication are absolutely central to psychiatric thought during the post-war period, even whilst they are anchored in, and stabilised by, much older concerns.

The emergence of social psychiatry, undergirded by the analytical tools of coping and stress, casts mental illness as a form of communication: attempted suicide as cry for help is an expression of, and a driving force behind, this turn to the social. The method here is to chart the rise and fall – between the late 1930s and early 1980s in Britain – of a particular set of techniques and institutional practices used to constitute and manage shifting forms of self-damaging behaviour. This does not presume an unproblematic or common-sense existence for these phenomena, but details the specific conditions in which meaning is produced. Overdosing as a cry for help is founded upon two principal innovations: institutional arrangements that focus consistent psychological scrutiny upon people presenting at general hospitals primarily for 'physical' injuries, and interventions that access and bring to relevance a credible 'social constellation' around the 'attempt'. This is predominantly based upon the evidence provided by psychiatric social workers; social work in general is central to the state's commitment to the management of social life. Self-cutting as tension regulation emerges first in inpatient populations, and is then projected onto different groups of people presenting at A&E. This focus on internal, individual tension reduction grows influential in a political context in which neo-liberalism stresses the virtues of individual self-reliance over collective provision. These are not simply strategies of interpretation or emphasis that enable a pre-existing overdose or self-cutting incident to become more visible or coherent. Practical and institutional arrangements and politically resonant sense-making produce this object in a fundamental sense. We shall see how, when and where this clinical and epidemiological object emerges and is consolidated into an increasingly common and explicable phenomenon.

Chapter 1 looks at an object under the name 'attempted suicide' prominent during the early twentieth century (1910s and 1920s). This is compared to one found in the late 1930s, in a mental-observation ward attached to a general hospital. This 1937 study marks the emergence of a distinctive psychological, psychosocial object. Chapter 2 assesses the significance of the Second World War (1939–45) and the subsequent founding of the NHS (1948) for this psychological concern, and subjects some of the work of Ivor Batchelor (1953–6) and Erwin Stengel (1952–8) to close reading, both in terms of their intellectual contents and institutional settings. Chapter 3 takes a close look at the Mental Health Act (1959) and the Suicide Act (1961), to see how various legal changes enable much broader governmental intervention focusing psychiatric attention upon physically injured patients, enabling the object to assume national (even 'epidemic') significance. Chapter 4 examines a government research unit on psychiatric epidemiology in Edinburgh, and on how the profession of psychiatric social work is vital in relating a hospital attendance to a social situation, calling the object 'self-poisoning'. Chapter 5 details the rise of a new form of 'self-harm' in Britain – self-cutting as a means of internal tension reduction – which surfaces during the 1960s (in both Britain and North America). The British literature on self-cutting is analysed, with the chief focus on how self-cutting emerges in inpatient settings and is gradually understood as motivated by internal tension, rather than analysed as a potentially contagious social phenomenon. This internal tension is then seen to differentiate self-cutting from self-poisoning; self-cutters are previously a barely remarked-upon minority in parasuicide studies at A&E departments. Self-poisoning then falls out of the spotlight somewhat, as the new behavioural phenomenon of self-cutting renders it ambiguous. The Conclusion describes the significance of this shift in broader terms: the displacement of overdosing by the prominence of self-cutting; a psychological object embedded in the social setting replaced by one focused upon internal, individual emotional struggles.

Concluding thoughts

The history of a particular psychiatric category is important because such categories are constitutive of human possibility. Hacking concludes that through these processes of (self) categorisation 'we are not only what we are[,] but what we might have been, and the possibilities for what we might have been are transformed'.[89] This history of cutting and overdosing in Britain can show how such coherences can come into use and how possibilities for identity are historically formed, linking the shifting analytical frameworks around self-harm to broader changes in cultural and political spheres.

This is important because, to quote Scott again, 'by exposing the illusion of the permanence or enduring truth of any particular knowledge ... [one] opens the way for change'.[90] The futures from which we are able to choose depend upon what we take to be the meanings of the past. If this position appears paralysing in stressing the multiplicity of the past, then it must also be able to demonstrate, in the words of Nikolas Rose, 'that no single future is written in our present'.[91] In this project, the people scrutinised, labelled, interviewed, referred, transferred, arrested, home-visited and otherwise assessed are made into and re-make these categories that render their behaviour somehow intelligible.

Finally, there are significant ethical implications for this kind of history. In showing how the meanings of the past and present are bound up with broader historical shifts, from social to internal, this book makes a point about the possibilities for change. For, if present meanings are the only valid ones, and history is merely an exercise in projecting those meanings backwards through time, history comes to naturalise the present, and offers nothing in the way of critical engagement. Instead, this book argues that the ways in which humans understand themselves and others are contingent, contextual and practical. The labels, and the kinds of labels, that we use have consequences that cannot be merely shrugged off by citing some eternal, intractable undercurrent, that validates (and is validated by) the imposition of current terms onto the past. Not only must we take responsibility for the descriptions that we use, it is incumbent upon us to be aware of how they fit into – and naturalise – broader transformations in thought and practice. The displacement of 'the social' (and with it much of the post-war welfare settlement) is a matter of great concern that this book, in its own small way, attempts to address. I am also concerned at the increasing reduction of human potential to biology and neurology in contemporary neuroscience, and the ways in which scientistic and neo-liberal business practices are being used to discipline and neuter the critical functions of higher education. This book takes as its method Foucault's notion of historical critique, which

> is not a matter of saying that things are not right as they are. It is a matter of pointing out on what kinds of assumptions, what kinds of familiar, unchallenged, unconsidered modes of thought the practices that we accept rest.[92]

Except where otherwise noted, this work is licensed under a Creative Commons Attribution 3.0 Unported License. To view a copy of this license, visit http://creativecommons.org/licenses/by/3.0/

1

Early Twentieth-Century Self-Harm: Cut Throats, General and Mental Medicine

At some point before five P.M. on 25 June 1914, in the small coastal town of Lowestoft, Suffolk, 59-year old Louisa Ashby cuts her own throat with a razor and lies down on her bed. Her eight-year-old granddaughter, Dora, discovers her covered in blood, and runs back downstairs to inform her mother that 'grandmother had cut her finger'.[1] Ashby is rushed to the nearby Lowestoft and North Suffolk Hospital, where, according to the East Suffolk Police:

> The [hospital] matron then requested that an officer should stay and take the sole charge and responsibility of the patient. I told her we could not do that, and that two of her sons were present [for this purpose], she said, 'They are no good, you brought her here and must take the sole charge of her, or take her away'.[2]

The matron accuses the police of 'not doing your duty...the woman has committed attempted murder [sic], and you should charge her...there is always this bother about cases brought here by the Police, and has been for years', and she even threatens to take Ashby and put her outside the hospital gates.[3] Ashby dies two days later. The dispute reaches the deputy chief constable who is unmoved, quoting East Suffolk Constabulary's general orders from 1902, to the effect that 'such patients are not in the custody of the police, [thus] he cannot take the responsibility of their safe custody'.[4] There is acknowledgement of ambiguity around the issue of responsibility, but there is one certainty: '[T]he police are responsible for ensuring that...at all events the offence shall not be repeated'.[5]

As this case is not read as a cry for help, and as it involves neither an overdose nor cutting of the arms, this might seem a strange place to start. The relevance of this case is that it shows how behaviour broadly conceived as self-destructive comes to the attention of hospitals (and more generally) in context-specific ways. Focusing upon how hospitals become concerned with self-harming behaviour before the 'overdose as cry for help' epidemic, between 1950 and 1980, sheds important light upon it. The idea of 'self-harm' as we presently understand it does not exist in 1914. The late-Victorian concern labelled 'self-mutilation' is significantly different, as it includes practices such as swallowing or inserting needles into oneself, self-castration, enucleation (eye removal) and eating rubbish, alongside the more familiar cutting, flesh-picking and self-biting. As Sarah Chaney clearly states: self-cutting 'is not emphasised in nineteenth-century writings'.[6] We shall also see later how the early 1950s communicative attempted suicide is distinct. The Ashby case is an example of what is called a 'would-be suicide', a concern that involves hospitals and police, as well as workhouse staff and coroners. This chapter offers an explanation as to why a cut-throat would-be suicide is a concern in the early twentieth century and why, towards the late 1930s, it might be displaced by the beginnings of a different kind of self-harm, haltingly conceptualised in more social, interpersonal terms.

A Home Office file at the National Archives documents a series of disputes between hospitals and police forces in England and Wales over patients like Ashby, thought to have attempted suicide and brought to hospital by police (suicide and attempted suicide are illegal in England and Wales until 1961). On a practical level these records exist due to a debate about who is responsible for taking custody of the 'would-be' suicide in the absence of a police charge, and whether the cost of watching these patients should be borne by the police.[7]

This financial dispute centres upon characteristics of 'renewal' and 'violence'. Broadly, renewal expresses a concern that the attempt will be repeated, usually at the first available opportunity, having failed the first time. Thus the attempt is cast as a genuine effort at ending life. Although the terms 'renewal' and 'repetition' are used interchangeably to describe this, renewal is preferred here to emphasise the difference between this concern and post-war usage. In the later period, 'repeated attempted suicide' indicates that a person resorts to an attempt at suicide at a number of different points, with each repetition considered distinct; subsequent efforts are not seen as trying to rectify the results of earlier suicidal episodes.[8] The second characteristic, 'violence', is more self-explanatory. However, in this context it is not always clear whether the violence is imagined as predominantly self-directed or directed towards others (in the former case it is largely indistinguishable from a renewal of the attempt). Violence and renewal are central because if patients are thought likely to renew their attempt or use violence then the Home Office considers that police are obliged to watch them, or to pay for civilian watchers to ensure that this does not occur. The obligation is thought to exist even if the person has not been charged with the common-law misdemeanour of 'attempted suicide' (and therefore is not formally in the custody of the police).

In this way, characteristics of this would-be suicide are bound up with context-specific economic concerns. Some police officers see much police time lost on behalf of 'nervous medical superintendents' who push for police to watch most cases; on the other hand hospital staff express resentment at the police bringing in cases that constitute a drain on voluntary hospital funds.[9] In the pre-NHS era, these are charitable funds, either an endowment from a wealthy person, or subscriptions and voluntary contributions from members of the public. Care at voluntary hospitals is considered 'better than the poor law, if one could get it', but this is bound up with being deemed worthy of charitable relief, or having a letter of recommendation from a subscriber or governor.[10]

Part of this financial dispute mutates into a therapeutic dispute with financial consequences. This concerns violence again, but also a new category of 'restraint'. At issue in the therapeutic dispute is whether the most significant aspect of attempted suicide is the somatic, physical, injury or the presumed underlying mental disorder. Ideas of renewal and violence, emphasised in the practical negotiations around police involvement, have another set of resonances with mental disorder through the presumed need for restraint. This aspect emerges most clearly at a 1922 inquest into the death of William Bardsley, a clerk from Stockport. Administrators and workers at a voluntary hospital turn Bardsley away, claiming that their hospital (and others like it) are unsuitable for attempted suicides because of the potential for violence, which is seen to require the restraining capabilities of mental therapeutics. The mental blocks of workhouse infirmaries (not asylums) are considered more appropriate. Bardsley is sent to a workhouse some distance away. Those in charge of the workhouse, the Poor Law Guardians, admit him. However, they do so without accepting the arguments of the voluntary hospital. Instead, they emphasise the somatic, surgical needs of his cut throat, claiming that the voluntary hospital is better equipped in that sense. Thus, although would-be suicides appear in the Home Office files due to a financial dispute, their emergence and significance is also related to a negotiation between the distinct therapeutic approaches of general and mental medicine. This division is constituted here between voluntary hospitals and workhouse infirmaries; mental hospitals refuse to take such patients until their physical injuries are stable. In addition, they also seem too geographically remote to be realistically considered in an emergency.

The respective positions of mental and general medicine shift in 1929–30, meaning that these debates over violence and restraint, over therapeutics and finance, recede somewhat. The archetype of the cut throat, and the violence and anxiety that surrounds it, is less relevant to the new context. Other methods and other readings begin to emerge, hand in hand with a sense of psychologically invested self-harm, where the goal of the behaviour is ambiguous – the beginnings of the concern with communicative self-harm. The Local Government Act 1929 abolishes the Poor Law, and the Mental Treatment Act 1930 broadens the scope for uncertified – so-called 'informal' – mental treatment. This brings mental and general medical therapeutics closer together, principally around the old workhouse mental blocks in former Poor Law infirmaries, now called mental observation wards in local authority hospitals. These wards are associated with mental illness and the use of restraint, but also as a diagnostic 'clearing station', a place where mental and general medicine interact, forming a distinctive field of visibility.

Finally, the work of Frederick Hopkins at a Liverpool observation ward can show how these combinations begin to make visible a communicative attempted suicide, through the opportunity for psychiatric scrutiny of patients presenting at hospital due to a physical injury. The methods most commonly reported here are coal gas, liquid corrosive and medication poisoning. Poisoning thus seems to resonate with the psychological ambiguity that becomes well-established in the 1960s. Hopkins's object emerges through an uneasy negotiation between the persistently separated approaches of general and mental medicine. General practitioner C.A.H. Watts recalls in 1966 that '[f]ew of us who qualified in the middle [nineteen-] thirties found ourselves equipped with any knowledge of psychiatry ... Medicine in those hospital days was almost completely an affair of organic diseases, and any psychiatric casualty was viewed as the usurper of a useful hospital bed – something to be removed with almost unseemly haste'.[11] The practice of mental and general medicine changes, as do the differences and negotiations between them. However, because this object is consistently seen as involving a physical element (the self-inflicted injury) and a mental element (anyone wanting to injure themselves must be mentally disordered in some way), it emerges reliably, though in a variety of ways, in a liminal space between these two regimes.

Renewal, responsibility and economics

In the case of Louisa Ashby, as noted, the Home Office decides that it is the police's responsibility to ensure that 'the offence shall not be repeated'. This concern with repetition or renewal also surfaces in a dispute over one Frederick Newman in Wiltshire in 1915. In this case the Home Office decides that although 'no charge of attempting suicide was made against him there was some risk of his repeating the attempt'.[12] The police are reluctant to charge a person with the offence of attempting suicide, because this involves taking responsibility for that person. However, hospitals consider such individuals as patients who need to be watched. Thus 'would-be suicides' emerge here (and are recorded as such) according to the quality of renewal.

This is inseparable from economic concerns. In 1914 the clerk of Lowestoft Hospital's Management Committee initiates the exchange over Ashby with the Home Office, emphasising 'the heavy expense which the Institution has to bear in the care of these Patients'.[13] The Home Office appears sympathetic to this point, advising the police that 'if as appears to be the case the Lowestoft Hospital is under private management and is supported entirely by voluntary contributions, the police have no very clear claim on the services of the staff in respect of cases brought there by them'. In addition: 'Mr. Reginald McKenna [Home Secretary] would be glad to know whether the question of making some contribution to the Hospital from Police funds has been considered'.[14]

Economics are also a concern for the police. In a 1923 letter from the Metropolitan Police to the British Hospitals Association, it is argued that due to economic necessity, the force has decided to stop performing duties that they believe 'cannot strictly be held to devolve upon them'. Thus they are 'unable to sanction the employment in all cases of Police Officers to watch would-be suicides'. However, they are 'prepared to do so in the comparatively few instances where the patient exhibits a desire to repeat the attempt, or is really violently disposed'.[15] In this way, 'would-be suicides' are characterised in terms of a specific debate around economics, to do with repetition and violence. In Liverpool in 1920, '[i]t is not suggested that the Police should supply watchers for all persons whom they may take to a hospital or infirmary after attempted suicide, but only that they should do so when there is reasonable ground for fearing that the attempt at suicide will be renewed or that other violence may be used'.[16] This economic concern brings out renewal and violence together, demonstrating that key qualities of this object of concern (its potential to be repeated and its violence) emerge directly as a function of a specific economic negotiation.

Violence and separated therapeutics

However, violence has a different salience in debates over whether would-be suicides should be treated in workhouse infirmaries or voluntary hospitals. In the early twentieth century, workhouse infirmaries are places where mental and general medical therapeutics co-exist to a greater extent than in many other institutions. The involvement of this boundary between therapeutic regimes in the emergence and persistence of attempted suicide runs throughout this book. However, it is constituted and negotiated in different ways in different contexts. In this particular discussion, the issue of appropriate care is brought to light in ways that still feed off the violence and economic concerns outlined above.

In 1907, a Home Office ruling on the correct place for these patients to be taken does not mention the facilities for treatment, but a more diffuse sense of the 'character' of certain cases. There is a legal obligation to admit emergencies to both workhouses and voluntary hospitals, but 'police should use discretion' when asking to admit cases to voluntary hospitals 'different in character from those which are ordinarily received there'.[17] This seems more to do with the type of case, rather than the character of the patient. It is possibly a continuance of what Geoffrey Rivett notes of early nineteenth-century voluntary hospital emergencies: 'Medical staff made a rapid assessment of the clinical priority of those attending, who were well aware that a judgment was also being made on whether they were fit objects of charitable relief'.[18] However, moralistic judgements bound up with charity could well continue to militate against admitting attempted suicide cases to voluntary hospitals in the early twentieth century. Whilst the Home Office clearly implies that attempted suicides are 'different in character' from other voluntary hospital cases, both workhouse infirmaries and voluntary hospitals are considered – from a legal standpoint in any case – equally valid.

In 1920 it emerges that the Liverpool police do not take would-be suicides to voluntary hospitals. They judge the workhouse infirmary especially suited for such cases due to the 'qualified persons' there. For this reason, extra expense on police watchers 'hardly seems justified'. This has turned from a diffuse and ambiguous concern about the type of cases admitted (with possible moral overtones) to a debate about therapeutic facilities – but still interwoven in a different way with economic questions. This feeds into an explicit statement about the potential violence of such cases: 'The official nurses [at workhouses] are expected to supervise mental patients, dangerous at times, when the risk of attack or injury to their attendants is much greater than that incurred through the care of suicidal persons whose violence would be probably only an attempt at further self-destruction'.[19] Thus facilities at the workhouse infirmary are implied to be appropriate for dealing with both the somatic consequences and the potentially dangerous 'mental' aspect of these cases. The Home Office response does not attempt to alter the terms of the debate. Whilst reiterating the position that violence is key in cases of attempted suicide, the argument also takes in the capabilities of ordinary hospital staff (i.e., not trained to deal with mental illness). The position is that '[t]he police should pay for watching of patients 'when there is reasonable ground for fearing that the attempt at suicide will be renewed or that other violence may be used and the ordinary hospital staff is insufficient to prevent it'.[20] The idea of a potentially violent would-be suicide is in a central position in an economic battle that is also fought around assessments of appropriate facilities.

It is unsurprising that would-be suicide is constituted on a specific continuum of violence when the whole administrative machinery by which such cases are looked after – and their care paid for – hinges upon assessments of that violence. But the debate about potential violence is also inextricably bound up with the question of how far an attempted suicide indicates mental illness.

A 'joy ride' between separate therapeutic regimes

The intimate relationship between assessments of violence and the suitability of general or mental therapeutics is clearly illustrated by a 1922 dispute at Ashton-under-Lyne, a small town between Manchester, Oldham and Stockport in the North-West of England. The inquest following a man's death causes enough of a stir to be covered by the London *Evening Standard* and the *Manchester Guardian*. On 27 January, William Bardsley, a clerk from Stockport, arrives at the District Infirmary, Ashton with a cut throat. He is refused admission and taken to the Lake (workhouse) Hospital, where he is admitted as an emergency, even though he is not from an area covered by that Poor Law Union. One result of the dispute is that the patient is ferried between institutions in search of treatment. At the inquest into his subsequent death it is observed that '[i]t is very hard to give a dying man a "joy ride" between hospital and hospital'.[21] This is a clear indication of the separation of one type of scrutiny from another, which is particularly problematic in emergency cases. The dispute over the appropriate care of attempted suicides is articulated in terms of 'attempted suicide as physical injury' (appropriate for voluntary hospitals) against 'attempted suicide as mental disorder' (appropriate for the mental block of Poor Law infirmaries).

The roots (and often the buildings themselves) of what become observation wards lie in these mental blocks of Poor Law infirmaries such as Lake Hospital. Hugh Freeman notes that Poor Law Union infirmaries are built during the 1860s to care for the increasing number of workhouse occupants who are 'ill or decrepit', and further, that 'most infirmaries had an observation unit or "mental block"', where cases are admitted and then either transferred to a mental hospital or discharged.[22] After the Lunacy Act of 1890, which 'consolidated previous legislation on emergency admission', observation wards are set up and 'mainly sited in Poor Law hospitals, and aimed to provide initial assessment of mental illness as a preliminary to admission to a mental hospital'.[23] St Francis's observation ward in South London, the source of much of the clinical material in Stengel's *Attempted Suicide* (1958), is part of the Constance Road Workhouse from 1895 until 1930, when the institution is renamed St Francis' Hospital.

At the start of the Ashton controversy, a letter is sent to the guardians of the Lake Hospital, explaining the (voluntary) district infirmary's position. Some time before the incident occurs, a pre-emptive letter is sent by the infirmary to the local police asking them 'not to send to the District Infirmary cases which they might have cause to consider were cases of attempted suicide'.[24] The extent to which this relies upon the attempted suicide being cast as a mentally ill rather than physically injured case is clear:

1. That it is a rule of the District Infirmary that persons of unsound mind should not be admitted as patients. 2. That most juries find that a person who commits suicide does so while temporarily insane. 3. That under a Home Office Regulation the Police are not now called upon to provide an Officer to watch over such cases where the patient is not under arrest. 4. To send such a person to an Infirmary like the District Infirmary, Ashton-under-Lyne, is liable to cause distress to other patients, and considerable dislocation and possible addition to the staff.[25]

In the four above points, mental state, police practice and financial cost ('addition to the staff') are woven together to cast would-be suicides as mental patients more suitable for the workhouse mental ward attendants.

The importance of appropriate staff/facilities is demonstrated by the coroner at the inquest, who invokes the concerns about violence, stating that 'he understood the Infirmary authorities could not take cases of suicide [*sic*] because they had not the necessary staff to deal with patients who might become violent'. A cut throat evinces a suicide attempt which, in turn implies violence. Thus, the facilities at the district infirmary claimed to be unsuitable, and they should not provide (or pay for) treatment.

The following exchange, reprinted in *The Reporter* newspaper, shows how seemingly exclusive mental and physical therapeutics become absolutely vital to the resolution of this case. Dr O'Connor, assistant medical superintendent at Lake (workhouse) Hospital argues that 'the patient should have been detained at the Infirmary where the staff had more experience of surgical cases, and was more accustomed to dealing with them'. He explicitly casts the case as one of somatic injury – a surgical question. The coroner responds that 'there were no male nurses at the Infirmary', which is incomprehensible – given the irrelevance of nurses of any gender to the propriety of surgical procedures – unless it is seen as bringing the argument back to a debate about restraint. H. Hall Daley (clerk to the guardians at Lake Hospital) clearly understands this as he replies that they do not have any male nurses either: 'We only have the mental ward attendants'. The coroner's reply explicitly positions attempted suicide as more mental than somatic before eliding this into a supposition of potential violence through the method of injury: 'Well, a case like this is treated more as a mental case. At the Infirmary I am told they don't receive cases where violence has been used'. Violence again emerges here explicitly as a function of a debate about appropriate hospital provision, across a psyche/soma split. However, O'Connor is not done and attempts to drag the case back onto somatic terrain, where the attempted suicide would be more suitable for the infirmary: 'in cases of haemorrhage it was essential that a person should be attended to as speedily as possible, and the Infirmary was equipped for that class of work'. Daley adds that 'the Infirmary, which largely existed for surgical cases, was better equipped to deal with that class of patient'.[26]

The negotiation of psyche and soma takes place across a divide between workhouses and general hospitals. These positions are not disputed, and Daley openly acknowledges the presence of mental nurses. The debate is pursued through a contest over whether the essence of a case of attempted suicide is mental or physical. The contested essence in this particular context enables violence to be consistently invoked. Thus the potential for violence emerges between therapeutic techniques. The *Manchester Guardian's* report emphasises the financial aspect over the therapeutic dispute.[27] However, rather than reduce the significance of the case to any one primary cause, it is useful to sketch out the arguments pursued in these different registers. The arguments that reach the Home Office are more likely to involve the spending of public money and the police, whereas those issues recede in a coroner's court where it is a question of establishing fault or not in a particular death. This becomes transposed onto the technical question of facilities (which is accepted by both parties) and the question of facilities best equipped to deal with violence. The point is to lay out a field of argument, structured by a specific mental/physical divide, where attempted suicide emerges.

Differences and similarities – rupture and continuity?

The characteristic of violence is almost totally absent from the post-1945 epidemic of attempted suicide. It might be argued that this is because 'self-poisoning' – the most visible method until the 1980s in Britain – is passive, and that cut throats used in the overwhelming majority of disputed cases here is an active and violent method. However, this book seeks to understand why certain methods emerge in certain contexts, in the course of specific debates. In a dispute involving police presence and the division between mental and general medicine, it is no wonder that violence and repetition come to the fore. Dealing with violence through restraint is seen as a key part of the job for both mental-ward attendants and the police (in their different ways), so the cases involving arguments for or against the presence of these professionals are likely to be described in those terms.

If we accept that there is no essential quality to any action independent of context, we can investigate how certain actions come to be classified as violent or passive or (self-) destructive. Because a cut throat usually involves a bladed object (considered in this context as more generally and immediately dangerous than a bottle of pills, for example), and because its repair seems to require the distinctly somatic specialism of surgery, this method seems most obviously to call for police involvement and also to straddle this somatic/psychiatric divide.

As for renewal, it might be argued that this has nothing to do with the context and that it is merely logical that a person who attempts to commit suicide and fails would be likely to renew the attempt, to complete the suicide. However, it is precisely a disruption of this logic that undergirds the post-war epidemic, arguing against ideas of attempted suicide as bungled or incompetent. The idea of repeated suicide attempts certainly emerges in the post-1945 discussions, but as noted above, this repetition is cast as a repeated response to social situations, an habitual coping mechanism, rather than as an immediate attempt to rectify the failure of the first attempt.

The violence largely disappears, and the repetition is fundamentally reconstituted. However, one aspect of these disputes flags up a subtle link between the attempted suicide of the 1920s and that of the late 1950s and 1960s – in addition to the idea that both emerge in the borderlands between mental and physical medicine. This concerns friends and relatives. Throughout the debate, the police consistently state that they are to employ watchers only until friends or relatives can be found to take charge. An order for East Suffolk Police from 1902 states that they will only pay for watchers 'where the person has no friends or relatives able to take care of him, or when such friends or relatives are unwilling to perform or pay for such a service'.[28] A Staffordshire Police order from 1904 states that '[i]t is always open, to friends or relations…to make such provision as they think fit for the care and medical treatment of these persons'.[29] In 1916 the Metropolitan Police commissioner states that the discretion over a charge for attempted suicide is 'based partly on the question whether the offender had any friends or relations willing to take charge of him'.[30] The consistent use of family and friends – and indeed the idea of watchers being a substitute for them – is a convenient administrative response to deal with legal ambiguity and supposedly nervous medical superintendents.

So whilst the notion of attempted suicide as cry for help has broad ancestry, it seems possible that the understanding of attempted suicide as primarily a communication with a social circle becomes more obvious if the first response of the police is to contact members of that social circle to come and watch over the attempter (a practice that does not totally disappear until 1961). This is not a case of one state of affairs being a 'prototype' of a later version of attempted suicide. During this period, ideas about the causes of psychological illness move away from concerns about heredity, the nervous system or brain lesions, and begin to focus more upon social relationships, emotional attachments and adequate adjustment (in infancy and adulthood), all things that place other people in a vitally important position in relation to a person's mental health. It is also the case that concerns about social issues – such as child guidance, marriage guidance and mental hygiene – emerge between the wars (see also Chapter 2). These concerns, which are decisively adopted by the state post-1945, feed into the self-evidence of the 'social setting' and its impacts. Thus, what begins as an administrative response to a suspected attempted suicide can obtain new intellectual resonance and salience. A practice rooted in the fear of renewal in general hospitals, and in a legally ambiguous situation, might also provide a basis (and an audience) for communicative self-harm.

Attempted suicide emerges at a point where confusion is keenly felt over the roles of the legal and medical professions in ministering to certain kinds of injuries (principally a cut throat) that require hospital treatment. Legal ambiguity, financial pressures (on both hospitals and police) and the separation of psychiatric and general medicine create a field of visibility for attempted suicide that emphasises renewal and violence as the two key characteristics. There is no sense of communicative self-harm in the Home Office and police files; instead there is a danger of repetition and a threat of violence (which does not consistently differentiate between a renewed attempt and violence towards others). Indeed, the fear of renewed attempt – which is the basis for employing a watcher – seems to at least imply some sort of earnest desire to kill oneself. The police contest that a watcher is always necessary, but there is no sense of a communicative demonstration. However, the consistent invocation of relatives or friends (the first port of call for watching those recovering from an attempt) might encourage the apparent self-evidence of an attempt at suicide performed as a communication to a social circle, a cry for help.

These disputes form a counterpoint to Stengel's lament in the late 1950s about the lack of machinery for the registration of attempted suicide. In the 1920s, would-be suicides emerge precisely because there is no single administrative, legal or medical body to assume responsibility for these cases. A more systematic process of recording emerges when the therapeutic regimes are not seen as a 'joy ride' away from each other. This begins to happen in the 1920s and 1930s, as the workhouse infirmaries are consolidated into local-authority hospitals and come to contain the potential for both mental and general medical scrutiny.

From workhouse infirmary to mental observation ward (1929–30)

The disputes in the 1910s and 1920s bring would-be suicide to light through a process of negotiation between the distinct therapeutic regimes of the voluntary hospital and the mental block of the workhouse, or Poor Law, infirmary. However, these blocks and observation wards come to form a much more complex space than suggested by the polemic pursued in the Ashton inquest. They become more prominent during the 1930s as mental observation wards. To sum up mental observation wards in early-to-mid-twentieth century Britain is difficult. Richard Mayou, founder and first chairman of the Section for Liaison Psychiatry at the Royal College of Psychiatrists, laments that '[l]ittle is known of how they operated'.[31] They vary widely in their functions and available resources, according to place and over time. These disclaimers aside, an interwar observation ward might cautiously be characterised as having two main functions: first, as a place for the initial assessment of psychological disorder with regard to mental-hospital admission; second, for the temporary care of cases deemed acute, disruptive or difficult – often with the implication that mental abnormality is behind such behaviour. This workhouse heritage is widely acknowledged in the literature produced in the early 1960s around general-hospital psychiatric units. In 1963, two clinicians working at St Clement's Hospital in London note that 'the observation wards [are] situated mainly in the poorer municipal hospitals or [former] Poor Law institutions of the great cities' of Britain.[32] In Pickstone's 1992 case study of general hospital psychiatry in Manchester, he mentions that 'the ex-workhouse mental blocks...afforded the opportunity for an alternative mode of development' for psychiatric practice not centred on the county asylums.[33]

The wards are transformed around 1929–30. First 'the Local Government Act [1929] placed the old Poor Law Hospitals under local authority control'.[34] In 1938 a report on London observation wards comments that the 'chief feature of the [1929] reorganisation of the observation wards in the Metropolitan area has been the concentration of these wards in six General Hospitals'.[35] The Act 'empowered the London County Council to appropriate to their health service any workhouses used for hospital purposes'. In addition to the 1929 Act, 'Section 19 of the Mental Treatment Act, 1930, allowed the use of these institutions for the detention of mental patients'.[36] Thus the wards are further entrenched into both general medical and mental therapeutics. Not only are the wards brought closer to general hospitals, they are assigned a role (initial assessment) under the Mental Treatment Act of 1930 on a national scale.

The 1930 Mental Treatment Act (or the preceding Royal Commission, 1924–6) is often the starting point for twentieth-century histories of the integration of general and mental medicine in Britain. Walter Symington Maclay, a key figure in post-war mental health policy, is a keen advocate of integration, attempting to 'bring psychiatry into the stream of the rest of medicine'.[37] When, in 1963, he lays out three crucial twentieth-century events for psychiatry, he begins with '1930, when the Mental Treatment Act for the first time allowed voluntary admissions to mental hospitals and development of outpatient departments on a national scale'.[38] Whilst he considers the Lunacy Act (1890) and Mental Deficiency Act (1913) important, the 1930 Act 'ushered in the era of mental disorder as an integral part of medicine'.[39] The Act's integrative impact is widely recognised. In *Social Science and Social Pathology* (1959), Barbara Wootton quotes the preceding Royal Commission's recommendation that the law should be changed so that 'the treatment of mental disorder should approximate as nearly [as possible] to the treatment of physical ailments'.[40] In Maclay's reading, especially, the story of twentieth-century psychiatric progress in general seems identical with the processes of integration between general and psychiatric medicine.

The act enables local authorities to establish psychiatric outpatient clinics, and treat patients without formal certification, integration that is also helped by local health authorities appropriating observation wards and consolidating them into general hospitals. It is not often made clear enough that observation wards constitute a key intersection between general hospitals and mental medicine. This perception is central due to the enduring association between observation wards and attempted suicide.

Observation wards: diagnostics and the contested nature of treatment

In 1937, the *Journal of Mental Science* publishes an article describing St Francis's observation ward. Attempted suicide appears here as a distinct object: there are '33 cases of attempted or threatened suicide' admitted under Section 20 of the Lunacy Act and '12 suicidal attempts' admitted by police officers.[41] No further comment is given; the attempted suicides are not seen as a special target for investigation, but they are a distinct entity. In the 1938 report on the six London County Council (LCC) observation wards (by Aubrey Lewis and Flora Calder), patients 'with suicidal tendencies' are counted among the groups 'peculiar to observation wards'.[42] Similarly, Frederick Hopkins of Smithdown Road Hospital, Liverpool, in 1943 claims that there are 'three fairly common reasons for admission for observation...attempted suicide, epilepsy, and G.P.I. [General Paralysis of the Insane]'.[43] Lewis and Calder note that these wards are 'somewhat isolated from the whole system of the mental health services'.[44] Positioned between psychic and somatic therapeutics, and significantly associated with attempted suicide, the observation ward's attributes in the field of security and restraint are key in associations with attempted suicide.

During the interwar period observation wards are intended to accommodate patients on a temporary basis, but this does not mean that they take voluntary patients. (The increasing levels of non-temporary elderly patients, stuck in observation wards because there are no suitable places for them to go, is a cause for considerable concern.) Patients are usually detained for an initial three days; before this period expires a magistrate is required to see the patient. Detention can then continue for a further 14 days.[45] After this combined period of 17 days, the patient is usually either sufficiently recovered to be discharged or needs to be transferred, whether voluntarily or involuntarily to a psychiatric hospital. This time is usually spent observing patients in order to diagnose them prior to disposition, but this process becomes augmented by a growing (though contested) treatment role.

During the 1930s '[o]bservation wards are still in their infancy so far as their developmental possibilities are concerned – in fact we are still in the process of deciding what their purpose should be'.[46] The diagnostic function seems agreed in the 1930s; there is significantly more uncertainty about what else might be attempted in observation wards. Treatment is at the centre of the changes. The Board of Control (the national body that until 1959 oversees and regulates mental treatment in England and Wales) is against this, arguing in 1935 that '[o]nce it has been established that a patient requires treatment for mental illness, no time should be lost in transferring him to the mental hospital, which in general is the only place able to provide the specialized experience and the therapeutic resources necessary for successful treatment'. The board further states: 'Every improvement of the observation wards increases the temptation to undertake active treatment, a practice quite inconsistent with the main purpose of such wards, which is the diagnosis of doubtful cases'.[47] The Board of Control is clear: mental treatment must take place in a mental hospital, and only there; observation wards are diagnostic clearing stations and gateways to the more specialised mental hospitals.

This effort to keep mental treatment solely within mental hospitals is undermined by the wards' agreed role in diagnostic clearing. In 1940 the impossibility of separating psychological investigation from treatment is explicitly stated: 'Investigation *is* treatment – as those who deal exclusively with psychoneuroses constantly emphasize'.[48] Such investigation is central to the wards, in their role as a diagnostic gateway: there is 'a tendency to regard them [observation wards] as psychiatric casualty-clearing stations'.[49] The military language of 'clearing station' is significant, given the established links between the First and Second World Wars and the proliferation of psychiatric techniques.[50] The term 'clearing-hospital' first appears (according to the *Oxford English Dictionary*) in the *Lancet* in 1914. The term 'clearing-station' (deemed equivalent) appears in 1915. The former term has a history before the First World War: an article entitled 'The Casualty Clearing Station' states in 1917: 'Prior to the present war, this unit was designated a "clearing hospital"; but the nomenclature was altered to "casualty clearing station" soon after the commencement of the present campaign [the First World War]'.[51] These clearing stations come to prominence during the 1914–18 war, but it is in the Second World War (1939–45) that frontline psychiatric treatment is carried out in them.

There is also a non-military parallel, seen as David Armstrong traces twentieth-century social medicine back to a tuberculosis dispensary described as 'a receiving house and a centre of diagnosis...a clearing house and a centre for observation...a treatment centre'.[52] The functions of diagnosis, treatment and observation all feature in debates around observation wards. Given Armstrong's compelling argument that the logic driving the practice of this dispensary is the same as that driving community-focussed, social medicine, the imminent emergence here of attempted suicide, similarly rooted in social environments and relationships, is illuminating.

Observation wards are clearly implicated in the negotiation between psychiatric and somatic therapeutics, and some are even treatment centres in the 1930s: '[I]n certain cases, active treatment...is to be encouraged, and that in fairness to the patient, it should be practised whilst the diagnosis of difficult cases is proceeding'.[53] As treatment is a more involved form of scrutiny or practice than simply diagnosis, the level of psychological scrutiny in these wards is – unevenly – increasing.

Lewis's and Calder's findings in 1938 are more in tune with the sharp differentiation desired by the Board of Control, stating that 'these observation units function largely, if not solely, as clearing stations'. They note that '[i]n none of the wards did we find any attempt at prolonged treatment of the patients'. The operative word here is 'prolonged'; they visit St Francis and quote the published article detailing its practices at length in their report.[54] It should not be forgotten – at the London wards explicitly – that psychiatrists who worked at the prestigious and world-leading Maudsley (psychiatric) Hospital also visited observation wards, especially the regular visits to St Francis' by Edward Mapother (superintendent of the Maudsley before the Second World War) and then Aubrey Lewis (as professor of psychiatry at the Maudsley-based Institute of Psychiatry).[55] These special circumstances at St Francis are acknowledged: 'Few observation wards in other counties have consultant psychiatrists, officers and staff experienced in mental diseases, and all prognostic aids'.[56] Lewis and Calder end the report with a clear response to the treatment debate: 'The fact we wish to urge is that the observation wards as organised at present cannot be said to cater for the treatment of large numbers of mild and early cases of mental illness that remain in the community'.[57] The potential link with 'the social' or 'community' emerges explicitly.

Finally, observation wards are significantly associated with practices of physical restraint, which has an impact upon the referral of patients considered dangerous (either to themselves or others), regardless of how often such techniques are used. The observation ward's association with such patients has a history: a *Lancet* editorial from the 1930s characterises observation wards as a place for 'acute and dangerous mental illness'.[58] In the late 1930s one of the functions of the St Francis Ward was 'to secure the safe custody of patients pending their admission' to a mental hospital.[59] This role persists after 1945. In 1954, Edinburgh consultant John Marshall argues that '[e]very general hospital group should have a psychiatric service with out-patient clinics, in-patient beds for suitable cases, and an observation unit for disturbed patients',[60] suggesting a significant controlling or restraining function. The potential for restraint and security at an observation ward makes it more likely for attempted suicide to become associated with such wards during this period, based upon the truism that attempted suicides are dangers to themselves.

To summarise, patients are compulsorily admitted to an observation ward for up to 17 days so that diagnosis can occur and the necessity for mental-hospital admission can be ascertained; formal treatment is discouraged, but is sometimes carried out, regardless. Thus, interwar observation wards can be characterised in terms of diagnosis, treatment and security. Their role in diagnostic clearing marks them out as a boundary space between therapeutic approaches, where mental treatment *slowly* becomes more acceptable. These 'mixed' clearing stations have an obscure but striking relationship with a more socially focussed psychological outlook, in both military and non-military terms. Attempted suicide continues to emerge in these places due to the coincidence of mental and somatic concerns, reinforced by the secure provisions around mental therapeutics.

This chapter ends with one of the earliest attempted suicide studies in England and Wales. Whilst Stengel's work at observation wards throughout the 1950s is acknowledged as central in the twentieth-century concern around attempted suicide (see Chapter 2) the first published study of attempted suicide to emerge after the 1929 reorganisations and abolition of the poor law in England and Wales appears in 1937, a study conducted by Frederick Hopkins at an observation ward in Liverpool. This clinical object is fundamentally linked to the diagnostics, mixed therapeutics and secure nature characteristic of these wards.

Frederick Hopkins and attempted suicide (1937, 1943)

Hopkins is a rather obscure figure with an interest in child guidance (co-authoring an article on parental loss with Muriel Barton Hall[61]); in 1968 a lecture series is established in his name.[62] His work is mentioned above, describing three of the most numerous classes of patient (including attempted suicide) that pass through his former workhouse observation ward (in two divisions of a general hospital) at Smithdown Road (Liverpool) during the Second World War. The link with child guidance is important, as it links Hopkins with a profession committed to social management, which is drawn into the welfare state after 1945. In his 1937 study, 'Attempted Suicide: An Investigation', he relates that these two divisions potentially receive 'all cases of attempted suicide occurring in Liverpool'.[63] The association of these special wards with attempted suicide is made explicit. It has already been noted that in 1920 Liverpool police judge the workhouse infirmary especially suited for attempted suicides.[64] This is clearly related to the secure nature; the majority of those 'whose mental condition or behaviour demands restraint and/ or supervision must be admitted to suitable institutional care' and the majority of these 'must in the first place go into a mental observation ward'.[65]

It is noted that the observation ward does not quite have the general medical facilities to deal with emergencies, but links with acute somatic care are maintained through transfer: 'Severe and urgent cases [of attempted suicide] may be admitted to the nearest hospital, but a large proportion of these, if they survive, are transferred [to the observation ward] when able to be moved'.[66] Even severe somatic emergencies make it to mental observation. As noted above, attempted suicide is one of three common reasons for admission. It is significant that the other two reasons – G.P.I. (since the establishment of the physical Wasserman test) and epilepsy are among the most securely somaticised mental disorders of the period. There is also a sense that G.P.I. patients and epileptics both have the potential to be disruptive and/or violent. These two illness categories perform a negotiation between psychic and somatic medicine that is very different to attempted suicide, thereby showing that there is nothing fixed or inevitable about such crossover.

As noted, the rise of treatment in observation wards heralds a more intense type of psychological scrutiny. However, the treatment role is highly ambiguous at Smithdown Road: 'In hospital, under conditions sheltered from ordinary life, they [patients] can take a more objective view. They are enabled to discuss and disentangle their mental complexities, and there is an opportunity for readjustment with relatives and associates'.[67] Hopkins is open about the therapeutic effects that occur in observation wards – social adjustment with friends and family – without actively carrying out treatment.

Similarly, the intensity of the scrutiny Hopkins brings to bear on the attempted suicide patients is unclear. His study is undertaken to find out which factors are most important in provoking an attempted suicide. He initially states that '[t]he material and social conditions are known or easily investigated, and relatives, friends, relieving officers, police and probation officers are usually available to provide information'. However, he then changes tack, conceding that '[s]uch an enquiry obviously entails a great deal of work in the detailed investigation of each patient, the interviewing of relatives, friends and other informants'. He reveals that in a 1930s observation ward, with limited opportunities for psychiatric scrutiny, it 'was decided to limit the number to 100 cases, taking 50 consecutive admissions of each sex' and that '[n]o effort is made to consider...its psychological mechanisms'. For Hopkins, 'a real and complete understanding of the causes for such action would necessitate so prolonged and detailed a study of the individual as is impossible in practice'. In remarkably explicit terms, Hopkins argues that a study of the 'psychological mechanisms' behind attempted suicide requires 'a great deal of work' and 'detailed investigation' – something that is just not possible in these wards at the time.[68]

This does not stop Hopkins from speculating about these psychological mechanisms and their significance, speculation that yields something rather similar to communicative self-harm in these observation wards. However, it is notable how cautious he is when describing it:

It might be contended, and with reason, that in investigating a consecutive series of cases admitted to hospital on account of attempted suicide, one may be dealing not solely with cases who have attempted self-destruction, but also with a proportion whose motive was essentially different, viz., to produce a similar effect in order to gain personal ends. That is to say, there may be cases whose actions are essentially hysterical, or comparable to the self-infliction of disabling wounds. A decision on this point, especially after the event, is always a difficult one.[69]

The transformations that are already happening in observation wards (having a consulting psychiatrist such as Hopkins on the wards, for example) bring the potential to re-evaluate attempted suicide.

Hopkins mentions a certain kind of poisoning: 'coal-gas poisoning is by far the most common method, in females accounting for nearly 70% of all suicides' as well as the most common method overall.[70] He sees poisoning in general as associated with predominantly demonstrative attempts:

> The small number of poisoning cases that it was found necessary to send to mental hospital compares in striking fashion with the large percentage of what might be called the more violent methods. ...It may be that in this [poisoning] group there are many whose attempt has been more of the nature of a demonstration than a serious attempt at suicide.[71]

However, Hopkins remains aware of his research limitations when appraising the stereotyped view 'that suicidal attempts by women are commonly of the demonstrative, attention-seeking kind, without real intent to terminate life'. He is cautious and equivocal about this, arguing that although such a view may or may not be justified, 'this investigation has shown that women are little less determined than are men'. Hopkins judges his research resources and opportunities too meagre to firmly establish a phenomenon or to generalise it. This is not to say that resources available for scrutiny (time, money, research assistants, etc.) correspond precisely to various characteristics of different research objects. However, some relationship does obtain between research objects and the level of scrutiny that produces them. The text quoted above seems at first a significant counterweight to the gender dynamic that appears so strongly in the textbooks, a dynamic that feminises attempted suicide. In fact, Hopkins has a gendered reason of his own: 'Impulsiveness, lack of knowledge and preparation result in fewer fatal endings to their [women's] attempts'.[72] Hopkins's gendering is achieved on the basis of impulsiveness and ignorance rather than on gendered intent (although he acknowledges that the 'intent' argument has been made).

He again mentions the effort that has gone into his series: not only why the patient decided to carry out the attempt but also any prior circumstances. One of his key findings here involves the term 'domestic stress', which

> is somewhat vague, but is meant to include such circumstances as deaths in the family, quarrels and disharmony on various accounts, such as religion, inconstancy, maintenance, etc. It is not surprising that the numbers under this heading should be comparatively large when the emotional relationships of family life have so many aspects. As might be expected, the effects were more frequent in women, because to women life as a rule is focused domestically.[73]

He has no doubt that the large number of cases concerning women aged twenty-five or younger (twice the number of men in this age group) is 'is due to the hazards of love affairs and of early married life, misfortunes in these circumstances bearing more hardly on the female'.[74] Thus a domestic-romantic social setting is projected from an observation ward, in order to explain an attempted suicide. This socially focused explanation is clearly linked to psychological notions of stress.

This domestic social constellation is focused upon the events immediately preceding the attempt, part of what Hopkins calls 'precipitating causes'. These include 'mental disorder' (where 'the immediate cause of the action was the abnormal state of the patient's mind'), as well as '[d]omestic stress', '[b]usiness or economic stress', '[a]lcohol' or '[a]matory disturbances'. However, these exist in a dynamic relationship with much longer-term 'conditioning causes', which 'include characteristics of personality showing definite deviation from the normal (or average), and physical states that were the primary cause of changes in the mental attitude'. These more long-term factors are considered inaccessible to this research project. However, Hopkins is clearly aware of their import – again through his work in child guidance.[75] This interplay between past and present factors, either in the social environment or the broader domains of aetiology, is investigated and reconfigured by various psychiatric workers during the 1950s and 1960s. Principally, the shift occurs between those emphasising the aetiological significance of childhood emotional trauma and those focussed upon current domestic stress and marital pathology.

Concluding thoughts

Hopkins's socially embedded object is very different to the financial disputes of police watching, in which rejected patients are ferried between institutions across significant distances. The referral arrangements at Smithdown Road mean that Hopkins is able to aggregate psychiatric evaluations of patients whose physical injuries require urgent somatic treatment in the first instance. The secure nature of the ward also encourages referral of attempted suicides, who have technically committed a crime as well as being thought dangerous to themselves. There is also the question of growing psychological scrutiny through treatment, at sites attached to general hospitals, although Hopkins's research resources are still rather meagre.

At the Ashton inquest the essence of attempted suicide as either psychological or somatic is debated, corresponding to therapeutic regimes so separate that they are a 'joy ride' apart. After the reorganisations of 1929–30 a different context obtains. Along with the secure nature of observation wards, the key contextual factor in attempted suicide is its position between the two distinct regimes of mental and general medicine. These are broadly contained in the mixed diagnostic/therapeutic environment of an observation ward, but their potential connection is also enhanced by referral practices mentioned briefly by Hopkins. The emergence of a socially embedded attempted suicide centrally concerns this secure and liminal therapeutic space. It helps to reconstitute attempted suicide as a new object for scrutiny. This liminality within general hospitals remains the focus in the next chapter, in the context of a radical extension of activity by the state in the arena of social work (especially child and marriage guidance) and socialised medicine (the NHS).

Except where otherwise noted, this work is licensed under a Creative Commons Attribution 3.0 Unported License. To view a copy of this license, visit http://creativecommons.org/licenses/by/3.0/

2
Communicative Self-Harm: War, NHS and Social Work

In 1944, Henderson's and Gillespie's *Textbook of Psychiatry* notes the 'remarkable progress that has occurred in psychiatry in recent years in the teeth of war conditions, and even, to a limited extent, because of them'.[1] The Second World War nurtures and catalyses a large number of reforms and innovations in the thought and practice of British psychiatry. Attending to the psychological casualties of the Second World War generates a huge number of interpersonally focused psychotherapeutic practices. The psychological significance of personal relationships, of adjustment to situations, of communication and social interaction become central to the linked aims of maintaining military and civilian morale on one hand, and returning psychological casualties to service as soon as possible on the other. The link between the social setting and psychological well-being is not generated by the war. However, the war does give an enormous boost to conceptions of what becomes known as the 'psychosocial'.

Of no less import is the post-war settlement, particularly the National Health Service (NHS). Its enormous significance impacts psychiatry in diverse ways. Most important here is inclusion of mental health within the comprehensive service, which enables closer co-operation and referral between the fields of mental and general medicine, vital for the visibility of communicative self-harm. NHS funding removes the financial burden of attempted suicide from voluntary hospitals, detailed in the previous chapter. This results in practically all cases presenting at hospitals to be admitted to general hospital casualty departments. The integration effected by the NHS means that these departments assume a coordinating function. Continuing as places for acute care, they become a gateway to the varied specialisms of hospital medicine (surgery, urology, etc.). Their positions as acute, non-specialist, diagnostic departments means that despite the removal of financial or therapeutic dispute, attempted suicide as communication does not emerge consistently here. There is no sustained psychological scrutiny or follow-up, both of which are necessary for this to materialise. Thus there are two parts to the increased emergence of attempted suicide: a path between different therapeutic regimes, or a space that can encompass them both, and the possibility for sustained, high-intensity psychiatric scrutiny to construct an environment necessary for communicative self-harm. This environment is crucial to the complex intent presumed behind the act, shifting it from the achievement of death and opening up communication as a possibility on a broad scale.

The scrutiny of environment is bound up with the rise of child guidance, and especially psychiatric social work. These emerge with the mental-hygiene movement during the interwar period. Jonathan Toms notes that an important strand of this movement was based on the insight that the mind 'was not atomistic and it couldn't be understood separately from its environment'.[2] Allied with the NHS, psychiatric social work provides a more consistent focus upon the environment and on the health of children. A short film about changes to health care in 1948 states that 'the local council will have a new duty to provide home nursing, health visiting, and home help services... maternity and child-welfare services will be improved'.[3] The NHS and social work, along with expanded welfare provision, bring the 'social environment' into renewed focus. Communicative self-harm emerges on a national scale thanks to the foundations laid by this settlement. It falls away when this provision is radically renegotiated in the 1980s, with the rise of neoliberal economics. Again, this relationship is not simply causal – in fact it is not really simple in any sense. However, the central idea here is that political and institutional contexts are fundamental to the emergence of clinical, psychiatric concerns: humans make sense of the world with the intellectual and practical resources that resonate with their larger context.

Concern about children is influentially expressed in the burgeoning popularity and influence of John Bowlby's theory of maternal attachment, with emphasis on the psychological importance of the family and on the connection between mental disorder and social problems such as crime and delinquency. This therapeutic approach underpins a pioneering series of attempted-suicide studies in the early fifties. These are carried out in Edinburgh between 1951 and 1955 in an observation ward with historical roots different to those of the workhouse mental block. This Ward for Incidental Delirium (known colloquially as Ward 3) has less focus on security and restraint and more of an entrenched somatic medical focus – specifically around poisoning. The studies carried out in Ward 3 are significant because their findings are underpinned by collaboration between a psychiatrist (Ivor Batchelor) and psychiatric social worker (PSW) (Margaret Napier). The presenting physical injury is transformed into a communicative symptom of a disordered social situation by the investigative practices emerging from this collaborative effort, such as home visiting and follow-up interviewing.

Alongside these studies are a number of contributions by Erwin Stengel, both by himself and in collaboration with a PSW (Nancy Cook) and a psychiatric registrar (Irving Kreeger), including the seminal *Attempted Suicide: Its Social Significance and Effects* (1958). The practice of referral to observation wards is prominent in Stengel's work, as is follow-up interviewing, showing how transfer between acute somatic care and psychological investigation is further developed by PSW practice. Attempted suicide is still significantly associated with observation wards. However, the NHS not only removes financial disputes but also facilitates movement between different therapeutic approaches, helping PSWs and psychiatrists to collaborate on this object, further transforming it into a consistent and credible expression of interpersonal disturbance.

Broader concerns about the young erupt in moral panics over Teddy Boys and rock 'n' roll during the 1950s, more famous landmarks of that decade's cultural history than is attempted suicide. However, these all focus upon the same demographic group: adolescents and young adults. Attempted suicide thus resonates with broader concerns about young people, deviance, delinquency and subcultures. In 1953 the Reverend Chad Varah establishes a service from his London vicarage for people 'in distress who need spiritual aid' and a '999 for the suicidal'. The *Daily Mirror* coins the term 'Telephone Good Samaritans' for the service and it sticks.[4] Concern about the mental, physical and moral state of young people, and about suicide, distress and despair circulate throughout the 1950s, a decade overshadowed on either side by the Second World War and the swinging sixties.

War, therapeutic communities and psychosocial practice

The Second World War provides impetus, resources and fertile soil for the study of the psychological significance of group dynamics and social contexts. Tavistock Clinic psychiatrist John Rawlings Rees is appointed

Consulting Psychiatrist to the Army early in the war. He argues that 'out of the peculiar conditions created by conflict and national effort, there seem to have come some things that are of value ... psychiatry has perhaps matured more as a result of war experience than it could have done in five years of peace'. Part of these so-called peculiar conditions is the notion that '[f]rom having a somewhat limited function, psychology became suddenly a weapon of war, a method by which the fighting force could be improved, the interests of the individual better served and the health of the community ... safeguarded'.[5] It seems that the particular demands of total war – chiefly for people to act in the interests of the collective – encourages this maturation or development of psychiatry to take a certain form. The fact that collective martial effort (total war) spawns a focus upon collective or group experiences and dynamics is not coincidental.

But the war is not the whole story. Tom Harrison provides a fine, lucid study of the Northfield experiments, perhaps the most famous wartime studies of the psychology of groups and group dynamics. He traces a sense of psychosocial awareness through the crowd theories of Gustave LeBon, Wilfred Trotter's ideas on herd instincts and William McDougall's concept of the group mind – ideas proposed in the late nineteenth and early twentieth centuries. He also mentions the ideas of Sigmund Freud, W.H.R. Rivers, Melanie Klein, Ronald Fairburn, Joshua Bierer and the field theory of Kurt Lewin. Finally, he mentions two prominent health and mental-hygiene experiments from the interwar period: the Hawkspur Experiment in Essex, and the Pioneer Health Centre in Peckham. This certainly seems like 'an intellectual primeval soup' that ferments towards group awareness.[6] However, tracing influences and precursors in a rush of names and conceptual shorthand can lead to confusion. Presented here is a brief appraisal of how the Emergency Medical Service enables the integration of psychological scrutiny into general hospital practice. Later in this chapter we shall see how it also leads to increased prestige for psychiatric social work. Thus, can we see how the war catalysed the development and acceptance of specific threads in the story of communicative self-harm.

The war is clearly seen to impact upon the integration of mental health perspectives into general hospitals. James M. Mackintosh, professor of preventive medicine at Glasgow writes during the war about how out of the emergency hospital service has developed 'a growing emphasis on the mental health aspect of general hospital treatment'. He includes psychiatric social work in this, adding that for the treatment of long-stay (surgical or medical) patients in general hospitals '[t]he psychiatrist and the mental health social worker should be in the background, ready to advise on cases of special difficulty'.[7]

War conditions are seen as having wide-ranging impacts on the functions of general hospitals. This involves consideration of social and psychological factors as well as physical ones.

> Since the beginning of the present war there has been steady although limited progress in the conception that the general hospital has a specific function in restoring the sick to health and working capacity. This involves early assessment not only of the patient's physical condition and ultimate prognosis, but also of his mental attitude, his family background, and his suitability for the work in which he was previously engaged, all psychological as well as industrial problems.[8]

Here we can see that the increase of psychological scrutiny in general hospitals in the early twentieth century (traced in the previous chapter through the Mental Treatment Act and observation wards) is developed and encouraged by the Emergency Medical Service in wartime.

Harrison notes that military life interacted with psychoanalysis and social theory, a triumvirate that he claims 'led inevitably to the experimentation with group therapy on a wide scale within the British and other armies'. Whilst contesting that this was in any way inevitable, we can agree that there is certainly a productive relationship between these three factors. He goes on to say that 'it became increasingly obvious, as the war progressed, that group therapy was a logical extension of army life. This, allied with the large number of men requiring help and the relatively few staff available, led inevitably to widespread experimentation with the new technology'.[9] If we again downplay the inevitability, Harrison here shows how the practical conditions of army life might make groups obvious in an intellectual sense, and the resource shortages make group therapy attractive in a much more mundane, but no less powerful way.

Harrison describes this focus upon social networks and group dynamics in wartime practice in terms of a discovery of pre-existing needs. He is committed to the insights of therapeutic communities as true (obvious, inevitable), rather than as emerging as a particular, historically specific perspective. Nevertheless, he argues that

> the exigencies of army life...provided the final link in the chain, and whether group therapy would have ever gained such recognition without this fillip is uncertain. Clearly, there were individuals

promoting this form of activity before the war; but they were largely operating in isolation and in a more or less charismatic manner. The war led to ordinary psychiatrists experimenting with these new ideas.[10]

From this we can see how wartime conditions interact with pre-existing ideas and practices, fuelling the development of these socially focused insights. Rather than being inevitable, they rely upon specific contexts in order to be able to emerge as increasingly obvious or self-evident.

The NHS and psychological scrutiny during the 1940s and 1950s

Building upon the Emergency Medical Service, the NHS brings different specialist outlooks into a new, more connected relationship with each other. In the case of psychiatry, the Board of Control (the government department responsible for mental health care until 1959) is brought into the NHS, having unsuccessfully pushed for a separate administrative mental health care structure.[11] Thus the potential for crossover between mental and general medicine is much more widely available than being simply focused upon observation wards. A new combination of specialisms brings about new clinical objects, and observation wards are well-placed to build upon this, playing a central role throughout the 1950s. The NHS is also the first step in broadening the new field, combining acute-physical and psychosocial visibility – on a national scale – in general hospital casualty departments. For various reasons, these departments cannot quite sustain this, but play an important role in the growing visibility of this phenomenon.

The establishment of the NHS is widely viewed as an important step in the integration of psychological and general medicine. The final chair of the Board of Control, Walter Maclay, and epidemiological psychiatrist, John Wing, both cast the founding of the NHS as an intermediate stage between separated and integrated mental and general medicine.[12] The end point of this process (for Maclay) is the Mental Health Act 1959, covered in Chapter 3. This integration impacts upon the visibility of attempted suicide.

In 1947, after the passing of the NHS Act but before the 'appointed day' of inauguration in 1948, clinicians at the Withington Hospital in Manchester relate the appointment of 'a visiting psychiatrist' allotted around twelve beds'. This non-observation-ward method of embedding psychiatric scrutiny in a general hospital setting has consequences for the visibility of 'attempted suicide': 'Seventeen patients were admitted after attempts at suicide by various methods, the largest group being six cases of barbiturate poisoning'. They are even more explicit about the changes in terms of visibility: 'Very many patients who would formerly have been treated only by physicians are now recognised as requiring psychological examination'.[13] However, this experiment is very small-scale.

By April 1950 in Manchester it is decided that to achieve progress in psychiatry, services should no longer be based around asylums, in direct conflict with recommendations from the local psychiatric specialists. John Pickstone argues that this is driven by the idea that services based in remote mental hospitals with peripheral general hospital clinics 'will only serve to divorce the diagnosis and treatment of mental disorders still further from the broad stream of general medicine'. Instead 'new psychiatry posts would be attached to district general hospitals'.[14] Thus in the early years of the NHS, integration is achieved by creating administrative structures that minimise the space between mental medicine and the general hospital. Of course, explicit attempts at crossover unavoidably reassert difference. This is exacerbated by the Board of Control; George Godber, chief medical officer between 1960 and 1973 recalls that that 'largely at the insistence of the Board of Control', all mental hospitals and mental deficiency hospitals had separate management committees. He claims that '[t]here was no reluctance locally to having mixed management groups – it was the Board of Control's influence'.[15]

A&E Under the NHS

Casualty departments are important under the NHS, as the reception (and sorting) centre for all emergencies, including attempted suicide. However, Henry Guly notes that '[b]etween 1948 and 1960 there was little of substance in the medical literature describing casualty services'. Guly notes that it is even argued that A&E does not qualify as a specialism at all due to its generalised role, covering emergency care of all kinds.[16] A&E is a particularly unfashionable area for doctors of the 1950s, and it remains over-stressed, understaffed and under-funded today.[17] In 1956 T.G. Lowden, a consulting surgeon working in Sunderland, writes a series of three articles in the *Lancet* entitled 'The Casualty Department' (following his book of the same name published the year before). He opens the series comparing casualty to a secretary's office, calling it a 'coordinating mechanism on the medical side', often performing administrative rather than strictly clinical work.[18] This coordinating role, a key part of the comprehensive service under the NHS, is the practical arrangement that removes the disputes over the appropriate place to take attempted suicides. For A&E to become the 'given' place to take an attempted suicide requires the NHS.

In Lowden's *The Casualty Department* (1955), attempted suicide is a distinct concern. He describes a coma patient sent in by her G.P., who regains consciousness on the way to hospital and shows no signs of illness in casualty. She is discharged home with a future G.P. appointment. However, later that evening she takes a large overdose of the same drugs and the casualty officer is criticised for not admitting the case. Whilst Lowden is sure that there is 'no reasonable basis for the criticism', this example shows that attempted suicide achieves visibility (and causes anxiety) in casualty because it is read as a genuine attempt to end life – an attempt that might be repeated more successfully at any time.[19] This concern is similar to concerns over renewal in the police watching disputes.

Thus, despite the integrative shift of the NHS, Lowden's position in the 1950s is both cautious and clear – the divide between mental and physical therapeutics remains central to his thinking. He argues that because of coroners' almost invariable reference to 'mental instability' in cases of suicide, '[a]ttempted suicide should therefore logically be an indication for psychiatric treatment ... and all such cases should be treated at a mental hospital, unless the medical or surgical condition is so great that general hospital admission is necessary'. The mental hospital is the most appropriate place for an attempted suicide, so long as medical or surgical treatment is unnecessary, a position that evinces a clear psychological/ general medical differentiation. He acknowledges that mental hospital admission is not often effected, so 'cases of attempted suicide who do not require admission for their organic lesions often call for a decision on disposal'. Again, attempted suicide is an issue due to the dual concerns of organic lesions and emotional states, the recurring poles of soma and psyche:

> Much depends upon the circumstances, and particularly the emotional state of the patient. Young girls who make a half-hearted attempt to commit suicide because they have misbehaved and missed a period may often be returned to the vigilance of their parents.[20]

Some small, highly gendered fragment of what becomes the attempted suicide stereotype emerges at a casualty department. Such a case is characterised as falling between therapeutic regimes: unsuitable for mental hospital admission and unsuitable for admission on account of any organic injuries. Thus, nothing much can be done, and the patient should be sent home. The therapeutic approaches are still too separate; different arrangements for psychiatric scrutiny are required in order to register a need for any kind of extended surveillance or investigation. Whilst the NHS is a key step in integrating therapeutic regimes, and A&E becomes the single site for all emergency admissions, a socially directed attempted suicide does not appear as a credible research object here. The scrutiny available at A&E is not sufficiently psychological or intensive to fabricate a credible social setting around the presentation of attempted suicide; the sorting of casualty seems to emphasise the separation of therapeutic regimes rather than bringing them together.

However, alongside A&E there is a continuing link between observation wards and attempted suicide under the NHS. In 1949, the above-mentioned Withington Hospital (Manchester) experiment shows how at first the nurses 'were anxious to get every attempted suicide out of the hospital and into the observation ward'.[21] The success of the experiment undercuts the nurses' attitude that the observation ward is the only place for attempted suicide, but their reported first reaction exposes the traditional association. Ivor Batchelor argues in 1955 that in the case of attempted suicide '[w]here possible, immediate admission to the mental observation ward of a general hospital is the ideal arrangement'.[22] Batchelor's observation ward studies are considered next.

Ward 3 of the Royal Infirmary of Edinburgh

Ivor R.C. Batchelor publishes eight articles on 'attempted suicide' between 1953 and 1955, based on clinical work at the Ward for Incidental Delirium (Ward 3) of the Royal Infirmary of Edinburgh. He serves as a neuropsychiatrist in the Royal Air Force Volunteer Reserve during the Second World War and subsequently joins the Royal Edinburgh Hospital under D.K. Henderson.[23] Henderson has been mentioned as co-author of an influential textbook, but he is much more significant than that. Professor of psychiatry at Edinburgh between 1932 and 1954, he is second only to Aubrey Lewis as an influential mentor to twentieth-century British research psychiatrists. It is said that Lewis used to refer to Henderson 'with a combination of sincerity and irony...as: "The most distinguished psychiatrist in the United Kingdom"'.[24] Batchelor remains at Edinburgh for nine years, leaving for Dundee in 1956, and in January 1958 takes part in a published discussion on the 'Legal Aspects of Suicidal Acts'.[25] Erwin Stengel argues that Batchelor is 'the leading psychiatric authority' on attempted suicide in Scotland.[26] He collaborates on three of the eight articles with Margaret B. Napier, senior PSW based at the Edinburgh Hospital for Nervous and Mental Disorders. These studies emphasise the role of so-called 'broken homes' and alcoholism in attempted suicide, the two foundations of the socially focused aetiology they construct. They are equivocal about the formal appeal character, doubting whether it is always present. They worry that overemphasising this point might lead to an underestimation of the danger involved.[27] Before these studies are analysed more closely, their national and institutional settings are described from two angles: the potential for crossover between psychological and general medicine, and the provision of high-intensity, environment-focused psychological scrutiny. These concerns, central to the analysis of observation wards in the previous chapter, remain vital here.

Suicide and attempted suicide are not crimes in Scotland, a situation described in more detail in Chapter 4, which focuses on a research unit at Ward 3. The lack of legal sanction in Scotland is regularly invoked in the late 1950s by those campaigning for decriminalisation south of the border (part of the growing post-war legal interest in suicide covered in Chapter 3). The documents produced in the lead-up to decriminalisation bring to light a standing arrangement in Scotland of much relevance. The Home Office enquires about Scottish hospital practices in 1958 and discover 'a standing rule that patients who have attempted suicide are seen by a psychiatrist whilst still under treatment'. The history of this rule is not given. However, the general situation in Scotland is described as 'neither clear nor altogether re-assuring'.[28] After the change in suicide law, the Department of Health for Scotland again states (in January 1962) that '[t]here are at present standing arrangements at Scottish Hospitals for the psychiatric examination of patients who have attempted suicide and have been taken to hospital because of their injuries'.[29] Thus there are established arrangements in Scotland for focusing some form of psychiatric scrutiny (presumably from visiting consultant psychiatrists) upon patients presenting at general hospitals and read as having attempted suicide. However, only one Scottish site appears to produce studies of this phenomenon during the 1950s.

An idiosyncratic, contested observation ward

During the early 1950s Ward 3 is under the administration of Senior Psychiatric Registrar James Kirkwood Slater. Neil Kessel and Norman Kreitman both acknowledge the centrality of this ward to their respective work on 'self-poisoning' and 'parasuicide' in the 1960s and 1970s. The ward facilitates consistent psychological scrutiny of patients presenting with a somatic injury. Kessel comments in 1965 that there are 'auspicious circumstances' for studying this particular subject in Edinburgh, because for 'many decades the Royal Infirmary has had an "incidental delirium" ward for patients who required overlapping general medical and psychiatric care'.[30] Kreitman recalls 'an excellent clinical service' and an 'ideal research base'.[31] The two parts of the transformation appear explicitly: overlapping therapeutic regimes and the possibility for high-intensity scrutiny (psychiatric research). The ward has some fame at Edinburgh's medical school, known among 'countless numbers' of graduates and called a 'unique and traditionally hallowed charge in the Royal Infirmary of Edinburgh'. Much of what follows is based upon an unpublished 1962 memorandum (most probably written by Slater) stored at the Lothian Health Board Archives in Edinburgh, the best history of the ward available.[32]

The ward begins the twentieth century as a place to house noisy or otherwise difficult medical patients, a provision then extended to those brought in by police (including alcoholics with delirium tremens). This is further extended, after 1918, with the admittance of prisoners in need of medical procedures. Finally, at some unspecified point, those in authority discover that Ward 3 is 'admirably suited to their difficulties about failed suicides and thus followed other forms of poisoning, including the accidental ones'.[33] (Note the elision of attempted suicide with poisoning.) These difficulties are therapeutic and practical rather than legal, as attempted suicide is not a crime in Scotland. The ward's purpose significantly fluctuates over the century, but still fits into the pattern of associating attempted suicide with observation wards.

The memo exhibits anxiety over the use of coercive measures, specifically locked doors: '[T]his ward alone in all our hospitals is under lock and key. The modern view resents this as an anachronism'. The short-hand of the 'modern view' includes the shift towards promoting equivalence between mental and general medicine. However, too close an equation with observation wards is rejected by Slater, who argues:

> No right thinking person would deny that a modern hospital must provide accommodation for psychiatric observation and in the absence of this the psychiatrists have consistently cast covetous glances at Ward 3, but equally their claims have been defeated by the vote of the consulting staff who have recognised that, while a special opinion is likely to be sought, not infrequently, yet, in the first instance, every single admission to this charge was a medical or

surgical problem and that the psychiatric opinion was needed if at all at a later stage.[34]

A number of things require comment in this long, dense sentence. Firstly, that Ward 3 is coveted by psychiatrists, who desire facilities for psychiatric observation. This implies that the ward must fulfil this function, at least in part. Slater resists these claims by asserting the primacy of non-psychological therapeutics (the claim that every single admission is a medical or surgical problem). He admits that psychiatric input is valuable in the appropriate place, and is anxious to stress that the current liaison/referral system works well: 'For many years a most happy arrangement along these lines has been in operation to mutual advantage'.[35]

Slater is most concerned to preserve the overall control that he believes would be ceded to psychiatrists were Ward 3 to become simply an observation ward. This fear emerges implicitly in his proposals to divide the ward 'into three easily identifiable categories' comprising a psychiatric and psychological observation unit, a poisons unit and a miscellaneous ward, including medical care of prisoners. He proposes a link between a psychological observation unit (under the sole responsibility of the professor of psychological medicine) and a poisons unit directed by a physician, assisted by the director of anaesthetics, the kidney unit and others.[36]

Even though observation wards are substantially mixed in their therapeutic capacities (mainly by association with general hospitals), the psychological aspect is seen by Slater as preeminent; their full title is of course mental observation wards. The differentiation of therapeutic regimes is clear, as he concedes full authority to the professor of psychological medicine over the hived-off observation ward section, and brings in some very somatic therapeutics for the poisons unit (which he sees as far more central to the identity of Ward 3) with anaesthetics and kidney specialists. He is anxious that the ward is not swallowed up by psychological medicine, and that the psychiatrists remain involved on a referral basis only. Indeed, he is explicit about psyche–soma separation, indicating that the observation unit and poisons unit are 'quite separate charges although inter-related'.[37] To borrow a phrase from Ian Hacking, 'this is claim staking with a vengeance'.[38]

Stengel and Kessel stake counter-claims from the psychiatric side. Kessel argues in 1962 that the poisoning unit at Ward 3 'serves as a psychiatric sorting and disposal unit for cases of attempted suicide far more effectively than the traditional English observation ward, which dares cater only for those who have not rendered themselves unconscious or hurt as a result of their actions'.[39] Whilst Kessel cedes the 'poisoning unit' name, his focus is on psychiatric sorting and disposal, which is complemented by somatic therapeutics. Stengel claims in 1963 that 'in Edinburgh [attempted suicides] are admitted to an observation ward where emergency services for resuscitation are available – which is not the rule in psychiatric observation wards elsewhere'.[40] The ward is envisaged primarily as a (psychiatric) observation ward, with somatic therapeutics attached, rather than a poisoning unit with psychological scrutiny available on demand. The uneasy co-existence of psychiatric and somatic therapeutics is exceptionally well illustrated. Slater's proposed reforms do not happen, and this productive tension between therapeutic regimes continues, enabling the transformations involved in attempted suicide as a communication.

In both Stengel's and Kessel's accounts, the Ward's somatic therapies provide opportunities to scrutinise patients arriving at hospital with somatic injuries. In an account from the 1980s, historian E.F. Catford highlights the extensive role of social workers in this scrutiny, claiming that they 'play an important role and may find it necessary to keep in touch with patients of the [Poisoning Treatment] Centre and their families for a long period'.[41] The connections between social workers, families and post-war psychiatry are extensive and significant.

Politics, PSWs, and child guidance

As well as the institutional base of Ward 3, Batchelor's and Napier's attempted-suicide studies are significantly influenced by and accessed through the practices of psychiatric social work. This professional group are exceptionally important in bringing the social setting to bear in various ways. The roots of PSWs lie in mental after-care and the child-guidance movement. Vicky Long shows that in the late nineteenth and early twentieth centuries 'the Mental After Care Association deployed lady volunteers to visit its charity cases in their homes or places of work to check on their progress and resolve any difficulties'.[42] Noël K. Hunnybun, senior PSW in the Children's Department at the Tavistock Institute, also mentions this association in his history of PSWs.[43] Jonathan Toms argues that there exist four organisations at the heart of the mental-hygiene movement in the interwar period: the Central Association for Mental Welfare, the National Council for Mental Hygiene, the Child Guidance Council and the Tavistock Clinic. All these groups, he claims: '[P]romoted social work as an important ancillary profession necessary for good mental hygiene. In particular they supported the creation of a profession called "psychiatric social work"'.[44]

John Stewart and Hunnybun both agree that the development of PSWs is intimately bound up with child guidance.[45] Hunnybun traces the profession back through concerns expressed in Cyril Burt's *The Young Delinquent* (1925), which emphasises 'the importance of studying the child in relation to his family and social background'.[46] Concerns with 'families' and 'social background' are absolutely crucial to PSWs (and to attempted suicide), and the profession emerges from a tangle of mental aftercare, mental hygiene and child guidance.

On an institutional level, the Tavistock Clinic's department for children opens in 1926 and the Commonwealth Fund of America finances the London Child Guidance and Training Centre, established in Islington, North London, in 1929. This same fund provides start-up money for the Association of Psychiatric Social Workers that year. Child guidance grows substantially during the interwar period. John Rawlings Rees is in no doubt about the significance of this for social-psychological perspectives. He claims that during the interwar period:

> Child psychiatry became established and never looked back; probably it is in fact the most important contribution to health that psychiatry has made in this century. The social worker and the psychologist began here to demonstrate how great a contribution they had to make ... we owe much of our growing interest in the sociological and psychological aspects of our work to children's psychiatric clinics.[47]

From 1936 John Bowlby works at the London Child Guidance Clinic. Whilst there he is 'strongly influenced by the psychiatric social workers' casework approach and theorisation of emotional relationships in the family'.[48] Bowlby's most influential concept is 'maternal deprivation', which locates the potential for psychopathology in mother–child attachments.[49]

This reconfigures the crux of the parent–child relationship away from the intricate fantasies, envies and anxieties of orthodox psychoanalysis, focusing on 'real life events': 'Where most psychoanalysts assume that neurotic symptoms originate from the patient's inner world of fantasy, Bowlby remained firmly convinced that traumatic events in real life were more significant – not only actual separation and loss, but also parental threats of abandonment and other cruelties'.[50] This constitutes a crucial emphasis on the social origin of psychopathology.

As well as the establishment of the Tavistock's Child Guidance and Training Centre, the year 1929 sees the London School of Economics establish the first PSW training course for social-science graduates. The universities of Edinburgh (1944), Manchester (1946) and Liverpool (1954) follow suit.[51] Prolific PSW Elizabeth Irvine notes that PSWs can join the local authority mental-health services after these are reorganised following the Mental Treatment Act 1930, and numbers rise from eight to twenty-six between 1951 and 1959. This 1950s movement from mental hospital to local authority provides 'an opportunity to return to the focus on the patient in his family which had been eroded in many mental hospitals'.[52] Felix Post – who conducts studies around the same time as Stengel (early 1950s) and on the same London ward – also becomes involved with the role of the family in mental illness, citing H.B. Richardson's *Patients Have Families* (1945) as a 'pioneer work'.[53]

The PSW training courses in Edinburgh are based on the Department of Social Studies, unlike those at Manchester and Liverpool, which are part of the respective Departments of Psychiatry. Even so, it can be assumed that the Meyerian influence of D.K. Henderson over psychological medicine at Edinburgh makes it a conducive place for PSWs to work. This enables them to flourish, for whilst '[l]ip service was paid to Adolf Meyer's more global picture ... only a minority of psychiatrists seemed to take this seriously in practice. [Those who did] were the best friends of the PSWs, and valued their support in demonstrating the ... tensions and conflicts in the family and social situation'.[54] PSWs are again intimately concerned with access to family and social conflicts in the aetiology and course of mental illness. Eileen Younghusband is perhaps the single most influential person in the field of social work in Britain in the twentieth century. In her two-volume retrospective of British Social Work published in 1978, she notes the 'complementary role' of social work in the treatment of mental disorder, stemming from wider acknowledgement during the 1950s of 'the profound influence which the family and social environment had on the well-being and social functioning of mentally disordered people'.[55] Ideas about 'the family' and 'the social' are of great importance.

As noted, the engagement of British psychiatry with the Second World War generates a huge number of interpersonally focused psychotherapeutic practices. It is also argued that a key factor in Bowlby's work is becoming influential – both in the mental-hygiene movement and upon government policy – is the onset of war.[56] The war reproduces, institutionalises and catalyses many of these interwar insights. Maxwell Jones, pioneer of the therapeutic community, states that '[t]he war years were my salvation', as his work at Mill Hill on effort syndrome provides the basis for his first such experiment in this kind of therapeutic organisation.[57] Rees relates in 1945 that it 'often occurred to me during this war how adequate a machine this child guidance team has been. Quite unconsciously the organisation of the War Office Selection Boards... has turned out to be on exactly parallel lines. Here also there is a team: a psychiatrist, a psychologist, and the regimental officer whose function is more sociological than military'.[58] Here child guidance, the team approach and the war are run together as a powerful innovation. Instead of simply treating symptoms in order to return men to the front lines, Wilfred Bion at Northfield sees the army psychiatrist's task in terms of social adjustment, an effort to 'produce self-respecting men socially adjusted to the community and therefore willing to accept its responsibilities whether in peace or war'.[59] Tom Main, describes 'therapeutic social fields' through which patients would progress on their journey back to adjustment and health.[60] The language of community, social field and adjustment pervades these wartime endeavours to treat mental disorder.

Kenneth Soddy's booklet, *Some Lessons of Wartime Psychiatry*, recommends that a 'psychiatric social service' be established to deal with mental disorder, mental deficiency and maladjustment.[61] During the war, James Mackintosh, professor of preventive medicine at Glasgow, argues that '[t]he expected result of this [wartime] work is that local authorities... will desire to place the whole scheme on a more permanent footing and make their own appointment of a psychiatric social worker'.[62] Invigorated and validated by the war, the concerns of (psychiatric) social work, centred upon the family, the child and adjustment to the social setting go from strength to strength as part of a broad political project in post-war Britain.[63] Influential studies from Aubrey Lewis's Social Psychiatry Research Unit by George Brown, Morris Carstairs, John Wing and others build from this position of strength, focusing upon the role of the family in the course and recovery rate of conditions such as schizophrenia.[64]

Nikolas Rose describes this post-war project in terms of 'minimizing social troubles and maximizing social efficiency' and notes that psychiatric social case work, through ideas about familial relations, is able to access and intervene upon 'the internal world of the home... in a new way'.[65] Mathew Thomson argues that social workers are seen during the 1950s and 1960s as 'shock troops' of a movement to spread psychological and psychiatric understandings of self and surroundings, with 'an ability to reach into the home'.[66] Eghigian, Killen and Leuenberger describe a post-war 'new wave of state interventionism... directed at women, children, and families'.[67] The goal of all this prescription, intervention, counselling, casework, psychological analysis and measurement is to produce what Rose has called the 'responsible autonomous family',[68] a nuclear, private, productive unit comprising well-adjusted and physically and psychologically healthy citizens. This is the 'social setting' with which 'self-poisoning as communication' corresponds. Jonathan Toms has recently complicated this picture, drawing out the tensions and contradictions in this view of the family and the authority in vested in it. He traces a shifting dialectic of family authority, always containing the seeds of its own disruption, from Samuel Tuke's moral treatment at The Retreat in York in the nineteenth century, to modern psychiatry, via the mental hygiene movement and 1960s anti-psychiatry.[69]

Governmental concern with increasing the number of social workers is noted by Younghusband in 1951, who points out that the Cope and the Mackintosh committees are considering 'the supply and demand, recruitment and training of almoners, and of psychiatric social workers and other social workers in the mental health service'.[70] She is famously associated with the Younghusband Report (1959),[71] which leads to the establishment of the National Institute for Social Work Training (1961) and the Council for Training in Social Work (1962).[72] Explicitly political intervention is also noted by Richard Titmuss in his lecture to the 1961 NAMH Annual Conference. He notes that '[n]umerous Royal Commissions and committees of enquiry have discovered in recent years the virtues of the normal social environment – or as near "normal" as possible'.[73] This is key in the wider project of constituting Rose's 'responsible autonomous family', where this family is 'bound into the language and evaluations of expertise at the very moment they are assured of their freedom and autonomy'.[74]

PSWs are an obvious expression of this psychologised turn towards 'the social' as well as being key instruments in the development and increasing ubiquity of such perspectives. In 1951 Aubrey Lewis claims that 'until comparatively recently explicit concern about these matters was rare ... Times have changed. The psychiatric social worker is an essential member of the mental hospital or clinic staff'.[75] Younghusband calls for a new type of social work with 'a social frame of reference, a fuller recognition of the complexity of human motivation and behaviour, and particularly of family and social interaction'.[76] It is startling just how far Younghusband's general description of developments during the 1950s maps onto the object of attempted suicide being tracked here, especially the complex motivation, and social frame of reference. Again, this effort – an intervention to manage, treat and regulate the social setting in targeted ways – stands in stark contrast to the shrivelled (or streamlined, depending upon your perspective) social concerns of the British state post-1980s, after privatisation and an enduring rhetoric of self-reliance (see Conclusion).

Observation wards, PSWs and the production of the 'social setting'

The potential for access to both psychiatric and general medical therapeutic approaches at observation wards (as well as a casual association with 'attempted suicide'), meshes with a broad turn to psychosocial explanations and interventions during the early post-war years in Britain. However, it is not simply that the mixed scrutiny of observation wards is complemented by the psychosocial turn, but that PSWs are increasingly attached to such wards. In 1937 it is noted that '[t]he social worker investigated the history of many of these [observation ward] cases, often interviewing friends or relatives in their own homes, so that a better idea of the domestic conditions could be obtained'. It is also claimed that observation wards 'have the closest contact with the relatives'.[77] This is a space where a vision of the family or domesticity is likely to be brought to relevance and prominence.[78] In 1940 the observation ward's 14-day period of detention is described as an opportunity to have the patient's history and background investigated by 'that essential member of the unit, the psychiatric social worker'.[79] Hunnybun includes the observation ward as a potential setting for PSWs working with adults, adding with some satisfaction that PSWs are gaining in prestige and wider recognition.[80]

The PSW contributions in Batchelor's and Napier's attempted-suicide studies are described as carrying out follow-up, collecting social data and obtaining data from the families.[81] The arrangements denoted by follow-up comprise

> personal re-examination of the patient, or by interviewing the nearest relative or other responsible and informed person. In six cases a psychiatric social worker in another part of the country made a home visit for us; in two cases we got a written report from the individual's general practitioner; and in two further cases a written account from another reliable informant.[82]

A significant proportion of follow-up is carried out through home visits. John Stewart emphasises 'the centrality of the home to child guidance and the part therein of the psychiatric social worker' during the interwar period, and that 'through the medium of the psychiatric social worker' child guidance becomes less focused upon the child as an individual, with more emphasis upon 'the child in its domestic setting'.[83] Indeed, sometimes '[s]ocial workers sought to visit the home even before a clinic visit'.[84] Bridget Yapp, co-author of *An Introduction to Child Guidance* (1945) with Mary Burbery and Edna Balint, claims that the 'child's difficulties cannot be understood without the fullest possible knowledge of the circumstances of his life, including the sort of home in which he lives'.[85] PSW Moya Woodside uses extensive home visiting when collaborating with psychiatrist Eliot Slater on *Patterns of Marriage* (1951) which investigates 'assortative mating' using hospitalised soldiers. Woodside is 'wholly responsible for the field-work. In nearly every case a visit is paid to the soldier's home'.[86]

The second practice – collecting social data or the social history – enables psychiatrists' reliable access to the social setting, and Stewart notes that '[p]sychiatrists appreciated such "social history"'.[87] In this, much weight is attached to 'unsatisfactory parent-child relationships in the first months and years of life', and 'the social and cultural background of the patient'.[88] The influence of mental hygiene and child guidance is clear. Finally, extended interaction with relatives is seen as significantly new in the 1950s. Irvine mentions a 'traditional concern with families', but also that '[t]his kind of work presented new technical problems. Social workers trained mainly for the individual interview…then had to deal, in conflicted family situations, with the anxieties and rivalries aroused in every member by an outsider's private contact with every other'.[89] Thus PSWs utilise new techniques when rendering the patient's social constellation, home and domestic background.

Looking at these practices and intellectual frameworks in a more abstract and analytical way, we can see how the presenting problem is subordinated to a social constellation – the problem is recast as a symptom of disordered interpersonal relationships. In 1949 John Bowlby argues that 'more and more clearly…the overt problem which is brought into the clinic in the person of the child is not the real problem; the problem which as a rule we need to solve is the tension between all the different members of the family'.[90] Toms has many examples of this shift in child guidance: Tavistock psychiatrist Dugmore Hunter writes in 1955 of children being forced into illness by parents avoiding their own problems, and psychiatrist Jack Kahn describes in 1957 the 'maladjustment funnelled into [a child] by the group tensions of the family'.[91] This kind of shift, from the presenting problem to the (supposed) real issues of domestic setting, family relationships and social psychopathology, is precisely the shift that underpins ideas of communicative self-harm.

Batchelor and Napier: therapeutic crossover, intensive scrutiny and John Bowlby

In Batchelor's and Napier's studies, the combination of observation-ward scrutiny and PSW practice is made meaningful through the conceptual apparatus of John Bowlby, which, as noted, roots adult mental disorder in real life (as opposed to symbolic/fantasy) traumatic experiences of loss and separation in infancy. The opportunities for psychiatric scrutiny of physically injured patients and for access to a social, interpersonal, domestic background, are guided by the concept that childhood emotional deprivations feed into present psychopathology. Batchelor and Napier explain the attempted suicide as a frustration reaction, largely rooted in a pathogenic broken home in childhood. The intent or purpose of the attempt is particularly complicated because this principal aetiological factor (the broken home) is in the distant past compared to the attempt. An emphasis on social history over social precipitants is evident, but there is significant awareness of the social repercussions of attempted suicide.

The key sample behind their studies is the 200 consecutive cases of attempted suicide admitted or transferred to Ward 3 between 1950 and 1952. (It is notable, given the idiosyncrasies discussed above, that Batchelor and Napier call Ward 3 an 'observation ward' without qualification.) This sample provides many sub-populations for analysis – such as elderly, psychopathic, or alcoholic patients, and those known to have attempted suicide more than once. Of most interest here are the two studies that use the entire sample. 'Broken Homes and Attempted Suicide' (1953) and 'The Sequelae and Short-Term Prognosis of Attempted Suicide' (1954) constitute an initial analysis and one-year follow-up, respectively.

The opportunity for mixed therapeutic scrutiny emerges in the claim – advanced with some pride – that every patient is 'thoroughly assessed from the psychiatric, physical, and social aspects' before discharge, and thus any decision is taken 'on the basis of considerable knowledge'. Their liberal discharge policy for these cases is cast as exceptional: 'It might well be unjustifiable to dispose similarly of a group of attempted suicides who had been more superficially examined'.[92] The necessity of all three assessment areas – psychiatric, physical and social – is repeated in 'Management and Prognosis of Suicidal Attempts in Old Age': 'the physician, psychiatrist and psychiatric social worker should collaborate'.[93] This shows that as well as the mixed psyche–soma scrutiny, the 'social' is just as important. They emphasise 'how necessary it is in cases of nervous and mental illness to understand and to treat the patient in his social context'.[94] The crucial point here is that Ward 3's provision of psychiatric and social scrutiny has the potential to transform the significance of a patient who arrives at hospital presenting with a physical injury. This injury is read as a consequence and symptom of past emotional deprivation.

Social constellations, broken homes and Bowlby

PSW input is most obvious in 'Sequelae' (an article predominantly concerned with follow-up) where the 'Social Reverberations of Suicidal Attempts' are charted. It is claimed that

> a small number, about 5% of the total group of 200, improved their social positions as a result of their suicidal attempts. If their acts were attempts to manipulate the environment in a direction favourable to themselves, they seemed to achieve that purpose ... A similar small proportion of the group worsened their positions.[95]

This is a present social context, the aftermath of the 'attempt'. Charting these reverberations (from clinical, hospital-based samples) is acknowledged to be difficult. They admit that only the most obvious or extreme consequences could be discovered, and that they 'know nothing of what had been for the meantime repressed successfully, but which may later have a traumatic influence'. They are, however, 'impressed by how frequently the suicidal attempt had made no great commotion in the family group'.[96] This is 'the social', accessed through interviews with relatives and families. A presenting physical injury is transformed into a psychosocial event through information provided (with some difficulty) by a PSW.

The notion of a present-centred appeal – with explicit acknowledgement of Stengel's first publication on the subject from the previous year (discussed below) – is downplayed. Batchelor and Napier do acknowledge that many patients bring attention to themselves through their actions, and gain treatment as a consequence. They understand such a present-centred appeal through a notion of temperament, claiming that this is most often the case for 'temperamentally unstable individuals chronically in conflict with their society'. Whether this temperamental instability is due to developmental issues or innate qualities is left unsaid, but its significance is downplayed: 'It is doubtful if it is an element in all suicidal attempts'.[97]

Batchelor and Napier subordinate present conditions or precipitants to the idea that a broken home in childhood is more significant. Throughout the articles it is repeatedly mentioned as a crucial factor. The opening of 'Broken Homes and Attempted Suicide' (1953) draws explicitly upon Bowlby to claim that the 'social and medical importance of "broken homes" in affecting adversely the mental health of the children nurtured in them is now widely recognized'.[98] They note that Bowlby's *Maternal Care and Mental Health* stresses 'the supreme importance of mother love in infancy and early years', emphasising that 'a broken home in the individual's childhood is aetiologically of considerable importance'.[99] However, they do not quote Bowlby's assertion (in the same WHO report) that 'the concept of the broken home is scientifically unsatisfactory and should be abandoned...In place of the concept of the broken home we need to put the concept of the disturbed parent–child relationship'.[100] Contrary to Bowlby's attempts to throw out the concept of the broken home, Batchelor and Napier seek instead to preserve and refine it, using broader samples allied with a precise definition.

They extend the concept of 'maternal deprivation':

> The traumatic effects of a lack of mother-love in childhood are nowadays everywhere recognized. Our findings also seem to emphasise the importance of a distortion or lack or absence of paternal influences in childhood. In a patriarchal society, the father is the figure in the home probably of chief importance...In investigations of the broken home situation there has been a tendency to lay almost exclusive emphasis on the role of the mother: the bias needs correcting.[101]

Whilst this assessment broadens the blame for the seeds of psychopathology in early life, it is no less gendered in itself. The paternal role is linked to wider society, an example or template. The mother remains the provider of love. Batchelor's and Napier's attempted-suicide pathology is still a pathology produced through a model of the home that is explicitly normative: 'We have used the term "broken home" as it is commonly used, to imply that the children in that home have been deprived of a normal life with their parents'.[102]

Childhood situations are deemed the most pivotal, and yet most difficult to access:

To assess emotional climates with regard to their normality or abnormality, to express in simple objective or qualitative terms such things as parental quarrelling or rejection and cruelty in parental attitudes, to eliminate the bias of not only the patient but also of his observer ... to give more than a very impressionistic opinion of a certain home in the retrospect of (usually) many years, is, of course, a most formidable task.[103]

Batchelor and Napier admit that 'evidence has almost certainly been missed' and that their tables of data cannot 'give a full statement of the complexity of the situations which were revealed' even though 'in every case relatives were also questioned'.[104] The questioning of relatives by the PSW is explicitly intended to uncover the past social constellation, but Batchelor and Napier admit that 'we have only the roughest clues as yet about how this factor [broken homes] operates'.[105]

Collaboration between psychiatrist and PSW provides the former with authoritative access to a realm of social information unavailable to Hopkins's observation ward in the late 1930s. But rather than simply document how broken homes are unearthed and emphasised through PSW enquiry, it is possible to see how visions of the social setting might be organised through these conceptual assumptions. This is most visible around statistics, as a considerable amount of effort is required to produce meaning when combining a set of numbers and a social constellation. At first, it appears that numbers are the problem in themselves. Batchelor and Napier state that the statistical tables in these articles cannot give 'a full statement of the complexity of the situations which were revealed ... no indication has been given of how some of these unfortunates were driven pathetically from pillar to post for their shelter'.[106] Statistics seem inadequate to display the social constellation. This is reiterated in 'Alcoholism and Attempted Suicide': 'These bare figures give some measure of the great frequency, but can give no picture of the quality, of disturbances in the childhood home-life of individuals'.[107] Numerical knowledge seems unsuitable for expressing childhood emotional deprivation.

Although psychosocial attempted suicide seems unsuited to statistical expression, these articles also show how Bowlby's ideas organise meaning out of complexity, despite the limitations of statistics. Whilst it is claimed that '[f]igures can, of course, indicate [things] only very crudely', they have meaning, nevertheless: They are, [however], sufficiently striking: parental alcoholism occurred in 38.1% of the cases, loss of the father in 33.3%, loss of the mother in 21.4%'.[108] The striking quality is sufficient to trump any crudeness. In another example, the concession that '[b]are, numerical data can give, of course, only a crude picture of family situations' appears with the qualifier that 'these data are at least factual'.[109] Even more explicitly, in 'Broken Homes', commitment to complexity is significantly organised by overarching ideas:

> To discuss in isolation the importance of broken homes in the aetiology of suicidal attempts, is to incur all the risks attendant on focussing attention upon a single aspect of a highly complicated situation. On the other hand, the figures presented in the tables above are so striking in many respects that to abstract this aspect of the problem seems justifiable.[110]

A Bowlbian conception of a broken home organises these numbers into meaning. Historian Joan Scott argues that statistics are involved in 'organizing perceptions of "experience"',[111] but here, Bowlby's conception of psychological development organises these statistics into significance: 'There seems, therefore, to be a particularly close relationship, which is psychologically understandable, between broken homes and suicidal trends'.[112] This is explicit evidence of what might be foregrounded under certain conceptual schemes, through what appears to be psychologically understandable: a past social environment anchored around a pathological broken home.

Information about social environments in the past, understood through ideas of pathological broken homes cannot be well-expressed in numerical form. The information is deemed too complex, too rich, too varied, even too emotionally charged (children 'driven pathetically from pillar to post'), to be expressed by numbers. However, these numbers still have meaning, because the same ideas that make these childhoods relevant organise the numbers so that they are 'psychologically understandable'. I am not arguing that Bowlby alone connects psychopathology to disruptions of nuclear, normative family units (they also resonate with Adolf Meyer's life-events, for example). However, the connections between PSWs, child guidance, explicit reference to Bowlby and visions of childhood emotional environments show the importance of PSW input to this reading of attempted suicide.

'Broken homes': aetiology and intent in the past

Whilst this reading of attempted suicide clearly feeds into the broader psychosocial political projects in a general sense, there is a PSW-influenced aspect of Batchelor's and Napier's work that is particularly relevant for studies of suicidal behaviour: the issue of intent. The detachment of intent from a simplistic wish to die is absolutely crucial in the creation of an interpersonal, psychosocial disturbance from a presenting physical injury.

The historical nature of the Bowlbian broken home complicates intent through notions of development. The significance of a broken home for healthy development is clearly described in Batchelor's 'Repeated Suicidal Attempts' (1954) – leading to a 'low frustration threshold' from a lack of socialisation and minimal training in tolerating setbacks.[113] Ideas of development and adaptation help to undergird a socially inflected 'attempted suicide'. Batchelor makes this claim: 'We may suppose that a broken home tends to render the individual less adaptable and, therefore, more vulnerable to the stresses of adult life and in particular ... personal relationships'.[114] Thus any present interpersonal social context is mediated by a lack of adaptability caused by a broken home.

In Bowlby's terms, these failures of adaptation are underpinned (at least in *Maternal Care and Mental Health*) by analogy with embryological development. He argues that 'pathological changes in the embryo's environment may cause faults of growth and development ... This is a finding of great importance, which, as will be seen, is exactly paralleled in psychology'. A second embryological analogy is deployed, linking the severity of developmental faults to the maturity of the tissue damaged; the earlier the damage, the more severe the consequences. For Bowlby, this constitutes a 'biological principle' that can connect 'far-reaching effects to certain emotional experiences occurring in the earliest phases of mental functioning'. He is almost protesting too much when he rounds off the argument by saying that these ideas, 'so far from being inherently improbable, are strictly in accord with biological principle'.[115] Bowlby's encounter with ethological methods of sense-making and the languages of stress and coping (what Rose calls 'an heretical amalgam of psychoanalysis and ethology'[116]) proceeds throughout the 1950s. The ethological influences are only published in a coherent theoretical position in 1958.[117] It is not just the changes in Bowlby's account of this link between childhood experiences and adult attempted suicide that complicate intent. Any such temporal link disrupts simplistic notions of intention, as these pivotal experiences are temporally distant or unconscious (or both).

What is important here is that the social setting's importance is rooted in the childhood history of the attempted suicide patient; this history impacts upon the present through a disruption of the individual's ability to adapt and cope with present situations. Bowlby describes this as 'unseen psychic scars...which may be reactivated and give rise to neurosis in later life'.[118] The social constellation most relevant to this conception of attempted suicide does not lie in the environment that immediately precipitates the attempt, but in the deferred pathological effects of a childhood broken home, effects which stunt the emotional development of the individual. The social setting figures as past impediment, not present precipitant.

Psychiatric social work brings an exceptionally high level of social and psychological scrutiny through interactions with families and relatives, making attempted suicide meaningful through a past pathology and a present maladjustment. It is a highly complex psychosocial object, made credible because such involved scrutiny can be focused routinely upon people brought to hospital presenting with a physical injury. Psychosocial aetiology and intent are fabricated around a presenting physical injury by high-intensity, psychosocial scrutiny. The idiosyncratic arrangements at Ward 3 mean that the potential for this object to emerge at multiple sites, on an 'epidemic scale' is limited.

Stengel and Cook: PSWs and a present-centred appeal

The work of Erwin Stengel and Nancy Cook at London observation wards is central to the phenomenon of socially embedded attempted suicide. The extent to which this work resonates with developments in general hospital psychiatry is less well-known. Richard Mayou shows how Stengel's and Cook's reading of attempted suicide and the association of psychiatry with general hospitals are intimately connected:

> [A]ttempted suicide has accounted for a substantial proportion of the cases referred in descriptions of [psychiatric] consultation services published since 1960. However, until the 1950s, hospital cases of attempted suicide were rarely seen by psychiatrists, and indeed, the clinical characteristics were not defined until the publication of a monograph by Stengel & Cook (1958).[119]

Attempted suicide and psychiatric expertise in general hospitals are inextricably linked, and this object is seen to emerge with Stengel and Cook. W.H. Trethowan's 1979 recollections bear out this transformation from somatic injury to psychological cry for help. He does not recall a single lecture on suicide when he was a medical student at Cambridge

University and then Guy's in the late thirties and forties,[120] but does remember that

> in the unsuccessful attempts – whether these ultimately proved fatal or not – it was the more immediate after effects which excited the greatest clinical interest – such as the cicatrisation [scarring or distortion of bodily tissue] which might follow corrosive poisoning, or dealing with the partial exsanguination [blood loss] and various surgical complications in those who had made more-or-less determined attempts to stab themselves or cut their throats.[121]

Trethowan attributes to Stengel's work (in the 1950s) the redefinition of such unsuccessful suicide attempts. Indeed, he claims that from his perspective in 1979 'attempts at suicide have become such a well-established form of communication between a person in distress and his environment that a satisfactory substitute is almost impossible to find'.[122] This shows how far the idea of communication has become entrenched – indeed, the shift from somatic to communicative concerns is explicitly linked to Stengel. The intellectual and practical labour underpinning this work is considered next.

Stengel studies medicine in Vienna in the 1920s, flees the Nazis in the late 1930s and enters Britain with the help of Ernest Jones and the British Psychoanalytical Society. He becomes one of the most successful and influential psychiatrists of the group escaping Central Europe in the 1930s (including Anna Freud, Willy Mayer-Gross and Joshua Bierer). He becomes a research fellow at the Crichton Royal Hospital in Edinburgh in 1942, director of research at the Graylingwell Hospital in Chichester in 1947, and reader in psychiatry at the Institute of Psychiatry (IoP) in 1949, as well as a consultant at the Maudsley. He takes the chair of psychiatry at Sheffield in 1957 and serves as the last president of the Medico-Psychological Association. Whilst his training (in 1920s Vienna) is unsurprisingly influenced by the psychoanalytical ideas, according to Aubrey Lewis's memorable phrase, Stengel is 'only singed by psychoanalysis'.[123]

Stengel publishes papers on 'Fugue States' (1941) and 'Pathological Wandering' (1943).[124] In 1950 he publishes a literature review on suicide, labelling fugue states 'symbolic suicidal acts'.[125] Thus his work begins to approach complex issues of suicide and intent. His first major clinical investigation of attempted suicide is based upon general hospital patients referred to mental observation wards in London. The cases that provide the basis for *Attempted Suicide* are split into five groups. Groups I and II are created using medical records from St Francis observation ward

(1946–7) and the Maudsley (1949–50), respectively. These records are used to identify cases and to attempt follow-up (the patients are interviewed by Kreeger, in his role as psychiatric research assistant, the relatives by Cook, the PSW). Group III consists of patients interviewed at St Francis by Stengel throughout 1953, soon after their attempt. Group IV reverts to the study of records, this time from a North London observation ward (St Pancras) for the same year (1953); these are compared with St Francis. Group V is accessed through an arrangement with Dulwich General Hospital, where every patient admitted there after a suicide attempt between 1951 and 1953 is psychiatrically assessed. (There is also 'Group S,' based on coroners' suicide statistics, which is kept separate and used as a basis for comparison and differentiation.)

St Francis's observation ward is the most important site, so a brief history is required. After the 1929 Local Government Act, St Francis becomes closely associated with Dulwich General Hospital; from 1948 they are under the same Hospital Management Committee (Camberwell).[126] Stengel's research project is funded by the Maudsley and Bethlem board of governors, and there are many connections between St Francis and the Maudsley, enabling access to high-intensity psychological scrutiny on a general hospital ward: Edward Mapother's, then Aubrey Lewis's, regular visits; W.H. Trethowan, who 'learned a lot' as a locum there when training at the Maudsley; Michael Shepherd recalls the 'old observation ward at St Francis Hospital with which I was associated for a long time'; Felix Post conducts studies there.[127] These arrangements and connections provide consistent psychological scrutiny from a world-leading centre of psychiatric research to a ward of a general hospital.

For Stengel, Cook and Kreeger, '[t]he self injury in most attempted suicides, however genuine, is insufficient to bring about death and the attempts are made in a setting which makes the intervention of others possible, probable, or even inevitable'. This repeated emphasis on the setting or environment is absolutely vital to the whole project. They argue for ambiguity in any intent, stating that '[w]e regard the *appeal character* of the suicidal attempt, which is usually unconscious, as one of its essential features'. They argue that 'if we think in terms of a social field we may say that those who attempt suicide show a tendency to remain within this field. In most attempted suicides we can discover an appeal to other human beings'.[128] This is a significant shift from Batchelor and Napier: a present-centred appeal underpinned by unconscious intent rather than a frustration reaction linked to childhood maladjustment. As the attempt is cast as a communication with the attempter's social circle, great pains are taken to document the circumstances of the attempt. Attempted suicide is rooted in the mixed therapeutics of observation wards, allied to PSW practice. In both Edinburgh and London, the different social constellations derive from and require intense PSW-enabled scrutiny.

Referral, therapeutic mixing and rising psychiatric scrutiny in 1950s observation wards

Whereas at Edinburgh's Ward 3 most patients are conveyed directly to that ward, a substantial proportion of attempted suicides in this study are referred to the London observation wards from general hospitals. In Group I (St Francis records, 1946–7), over half of the attempted suicides reach the observation ward, having been transferred from one of 16 local hospitals. In Group III (St Francis's patients interviewed by Stengel in 1953), over two thirds of attempted suicide patients are referred from other hospitals, rising to over 70% in the final observation ward group (Group IV from St Pancras). Combining all three observation ward groups, just over two thirds of attempted suicide admissions are transfers from other hospitals. This dwarfs the other methods of registering (by police and duly authorised officer). Stengel notes that the majority of attempted suicides are referred from other hospitals, something which is not the case for other kinds of observation-ward patient.[129] It is clear that consistent movement from a place of general medical therapeutics to a separate space with potential psychiatric scrutiny underpins the research. Whilst Batchelor and Napier rely on transformations enabled by mixed therapeutics, Stengel and Cook rely on a different crossover: established, well-used channels of referral. Hospital–observation ward referral is also crucial because there exists no central collection agency recording attempted suicide. Through referral, records of these attempts – which would have otherwise remained disparate – can form the basis of a research object.

In *Mental Illness in London* (1959) Vera Norris argues that the Board of Control's negativity towards observation units during the 1930s stems from the fact that many of the units are at that point situated in unsuitable public assistance hospitals, staffed by people without psychiatric experience.[130] Donal Early, surveying 15 years' change in observation-ward use in Bristol, notes that it was only the inauguration of the NHS in 1948 that prompted the provision of psychiatric cover to the ward.[131] Psychiatric scrutiny is increasingly provided for observation wards from then on, but in most cases it is judged to be at a low level. In 1954 John Marshall describes co-operation between psychiatry and general medicine as 'sadly lacking', and J.B.S. Lewis (superintendent of St Bernard's (Mental) Hospital, in Southall, Middlesex) labels observation wards as 'the weakest link in the administrative set-up for the mentally sick' because they are often run by clinicians without significant psychiatric expertise.[132] Despite this, during the 1950s there is a slow increase in psychiatric scrutiny on these wards. This increase should not be overstated, as even in the later 1950s, Norris observes that 'the primary function of these units is reception and diagnosis' and it is argued by others in 1961 that St Francis's ward 'preserves its traditional role of diagnosis and disposal'.[133]

However, *Attempted Suicide* is not solely based upon observation wards. The 76 patients in Group V are seen by a different arrangement at Dulwich General Hospital. General hospital psychiatry outside of observation wards is hugely uneven in this period. After leaving the Maudsley in the late 1940s, psychiatrist Max Hamilton joins University College Hospital (UCH), where: 'At first, they didn't know what to do with me. After a while, I managed to establish a job in liaison psychiatry...word got around that somebody was available'.[134]

Between 1951 and 1953 a special procedure is put into place at Dulwich to enable psychiatric scrutiny: 'It was arranged that during the period under survey every admission for attempted suicide should be seen by the psychiatrist in the team [Stengel]'. But, '[i]t is possible that sometimes he was not consulted...This applies particularly to patients admitted to the surgical department'.[135] Not only do Stengel and colleagues have to arrange to see the attempted-suicide patients, anxiety remains that patients might escape psychiatric scrutiny. Something similar is noted during a discussion of a five-year study of psychiatric referrals at Guy's Hospital in 1962. It is claimed that 'there is nothing new or unexpected in the observation that physicians call for psychiatric consultation more often than surgeons'. This is attributed to physicians' greater interest in psychological factors and surgeons' greater tolerance of mental symptoms.[136] Thus within a hospital – between the specialisms considered inside the label 'general medicine' – different regimes of referral and different professional identities complicate the constitution of any clinical object.

This state of affairs potentially blocks psychiatric attention from some of the more severely injured patients – for example, those who require surgery rather than first aid. Put another way, less gravely injured patients have more chance of obtaining psychiatric attention when brought to a general hospital under this arrangement. Equally, recalling Lowden's observations earlier in the chapter, such patients might be sent home from A&E. The potential for more seriously injured patients to escape Stengel's scrutiny has consequences for his ideas about demonstrative or appeal-based attempted suicide. Referral is a vital practice that bridges therapeutic regimes, but not without complexities and constraints.

Psychiatric resources, intensities of scrutiny and PSWs

The transformations that underpin Stengel's production of socially embedded attempted suicide are broached in the discussions of Hopkins's and Batchelor's and Napier's studies: principally that attempted suicide needs significantly mixed therapeutics and much intellectual and practical work for the transformation from a physical injury to a psychosocial communication. In Stengel's work, the present-centred social constellation around the attempt is indivisible from the intent presumed behind it. Intent to appeal cannot exist without some idea of a recipient. This contrasts with Batchelor's and Napier's analysis, where a broken home in the past impacts upon present abilities to tolerate frustration. A frustration reaction does not require the presence of recipients or observers, but remains rooted in a past, pathological environment.

Various practices, including follow-up and on-ward interviews (as opposed to simply the use of ward records), are required in order for Stengel and colleagues to make the observation-ward material yield up the communicative articulation of attempted suicide. There are three distinct sets of scrutinising practices: observation-ward records only, observation-ward records and PSW follow-up, and interviews with a psychiatrist on the observation ward. Observation-ward records alone constitute a low form of scrutiny. In the St Pancras observation ward, patients in 1953 (Group IV) are not interviewed. Early in the text it is claimed that the intent behind the action will form a key part of the discussion of Groups III and IV. However, this does not materialise for the St Pancras sample: 'Dangerousness and intent could not be assessed' because the patients are not interviewed by the researchers. For the same reason 'the social constellation at the time of the act could not be established'.[137] (The chapter on St Pancras does not fill three pages.)

Using observation-ward records and Cook's PSW follow-up allows a little more of the social setting to be fabricated around the attempt. In Group I, as well as sifting through ward records, the patients, their relatives, friends, and even employers are interviewed, subject to patient consent. The patients are mostly interviewed by Kreeger and the relatives by Cook. The interview schedules are reproduced in the text; they emphasise questions on matters presumably not found consistently in hospital records and case notes. For example, items on the psychiatrist's schedule for patients include: '[m]arked parental discord or other abnormal environmental stresses or relationships in childhood'.[138] Such questioning performs clear intellectual work, bringing patient history into a relationship with the suicidal attempt and opening up similarities with Batchelor's and Napier's work. However, the focus of the questioning is an exceptionally meticulous attempt to chart the present social environment through repercussions, a clear indication of their importance, and what is needed to achieve its prominence:

Changes in patient's human relationships and environment since attempt. The patient's views on the rôle of the attempt in bringing about changes in (a) social adjustment, (b) work and financial circumstances, (c) emotional adjustment, (d) sexual and marital adjustment – change in status, further children, etc., (e) change in mode of life of members of his family or friends.[139]

The PSW's schedule (for relatives) contains clear emphasis on the patient's relationships with other family members. The very existence of a schedule explicitly for relatives constitutes a research practice designed to produce an idea of interpersonal relationships related to an attempt at suicide. Most of these informants are seen 'in their own homes, as visits were regarded as essential for full information'.[140] Thus, the research object is produced from more intense scrutiny than the normal records can provide. This is acknowledged as vastly time-consuming in 1952 (in the write-up of the preliminary study, which features as Group I in the book), to the extent that Stengel is not surprised that the resources for this kind of study have not been previously available:

Only a small proportion of patients were in a mental hospital at the time of the follow-up. The rest had to be traced and their co-operation and that of their relatives had to be won. They proved a very elusive group and we came to understand why such a follow-up had never been carried out before in this country. I wish to pay tribute to my co-workers who overcame difficulties which often appeared insurmountable.[141]

That these patients have not been admitted to a mental hospital is part of the reason they are considered so difficult to trace. As models of psychiatric provision move away from mental hospitals, the techniques used to gather social, biographical and follow-up information around mental illness must change. It becomes clearer why Frederick Hopkins cannot produce such an interpersonal object in the late 1930s.

The 'Results of the follow-up' section contains substantial examples illustrating the social effects of the suicidal attempt. These include sub-sections such as 'Removal from the scene of conflict' and 'Changes in human relations and in modes of life'.[142] The first case study under the latter heading reads thus:

> Mrs. F.I., born 1910, was unhappily married...They separated in 1944...Soon after she learnt of her impending divorce, her lover told her that he did not intend to leave his family...She became acutely depressed and tried to poison herself with aspirin.... Three months after the suicidal attempt she resumed work. Her lover left his family after all and at the time of the follow-up six years after her suicidal attempt they were living together and both declared that they were thoroughly happy. She thought that her suicidal attempt had 'brought him to his senses'. Her family, who had been against this relationship had become reconciled...The suicidal attempt here contributed to the solution of a conflict.[143]

The attempted suicide is given meaning, but not as a symptom of a depressive illness, childhood deprivation or other psychiatric abnormality. Through follow-up, the attempt is given a social, communicative and instrumental meaning. A specific practical arrangement enables the presenting physical injury to be re-described as a communication.

The most intense scrutiny involves Stengel interviewing patients at St Francis in 1953 (Group III). He again claims that a 'number of aspects of attempted suicide cannot be satisfactorily studied months or years after the event. Some [of these aspects] have been investigated in this series, all of whom were interviewed...shortly after their admission'.[144] Thus the highest level of scrutiny achieved in this study involves a research psychiatrist interviewing patients soon after admission, with the investigation of the social element in attempted suicide as the purpose of the interview (allied with PSW follow-up). For a truly satisfactory clinical object, embedded within a social context, the on-ward interview is necessary. The potential for such a high level of psychiatric scrutiny is simply not available in observation ward of the interwar period.

The reconstruction of intent here is used to downplay the significance of somatic in favour of psychological consequences. Case studies illustrate this and contain frequent references to a social environment that modifies assessments of (physical) seriousness. For example, a woman who had taken a large dose of sleeping tablets and then 'called her sister with whom she was staying and told her what she had done ... Her attempt was graded as *absolutely dangerous, with only slight intent.* Had her sister not been available the attempt would probably have proved fatal'. A woman whose husband had been unfaithful 'took 100 tablets of codein-phenacetin compound when alone at home but knew that her son would come soon and she expected that he would find her alive ... The attempt was graded as *relatively dangerous, with slight intent'.*[145]

In light of all this effort, *Attempted Suicide's* most quoted passage takes on a different hue:

> There is a *social* element in the pattern of most suicidal attempts. Once we look out for the element we find it without difficulty in most cases ... If we think in terms of a social field we may say that those who attempt suicide show a tendency to remain within this field. In most attempted suicides we can discover an appeal to other human beings.[146]

The idea of looking for the social element, the intellectual move to think in terms of a social field, the discovery of an appeal: all these are dependent upon specific research practices. The social field is produced through them – finding relatives years after an event, sending letters asking for an interview, asking permission to speak to the former patient. It is quite a practical achievement to produce a credible social, interpersonal space around the paper record of an attempted suicide. Observation-ward records are useful, and follow-up is more useful still, but on-site interviews with the senior research psychiatrist are indispensable to a present-centred social constellation in which to position the suicide attempt in observation wards.

Concluding thoughts

Insights about the significance of social groups and social relationships to mental health and disorder are catalysed by the Second World War. Interaction between psychological and general medical scrutiny is strengthened by the inauguration of the NHS and the inclusion of mental health in the comprehensive service. This is the post-war social settlement, the welfare state and social support networks that are later rolled back by the neoliberalism of the 1980s. Attempted suicide emerges with greater regularity in mental observation wards in these socially focused times. These wards exist uneasily between separate therapeutic approaches, and the increased psychological and psychosocial scrutiny in them is of the highest importance for this new reading of attempted suicide. In this chapter it is shown that when crossover occurs – through mixed therapeutics, referral, or both – the scrutiny must be intense. Much of this intensity is provided by the follow-up practices and intellectual frameworks of psychiatric social work and child guidance, informed by a psychosocial focus that emerges energised from the war. It is also institutionalised by the post-war welfare settlement. The political will to intervene in, manage and treat the social setting feeds into and feeds off this psychological object. Whether the social constellation is fabricated around deprivations projected into childhood, or through a complicated, largely unconscious, appeal to a present social circle, it is highly labour-intensive.

The following chapter describes how crossover between psychological and general medicine is given publicity and impetus by the Mental Health Act 1959 and the Suicide Act 1961. The 1959 Act represents a peak in efforts to integrate psychiatric and somatic therapeutics – to which the 1961 Act is connected – through concerns about psychiatric scrutiny at A&E departments. As this latter act decriminalises attempted suicide, it alters formal NHS responsibilities for those considered to have performed that act. This impetus transforms attempted suicide from something of an observation-ward curiosity to a national epidemic. This has little to do with ideas of supposed 'actual' incidence. It has much more to do with the ways in which institutions and practices produce, maintain and expand new fields of scrutiny populated with socially embedded psychological objects.

Except where otherwise noted, this work is licensed under a Creative Commons Attribution 3.0 Unported License. To view a copy of this license, visit http://creativecommons.org/licenses/by/3.0/

3

Self-Harm Becomes Epidemic: Mental Health (1959) and Suicide (1961) Acts

At the end of Stengel's 1952 paper, 'Enquiries into attempted suicide', he speculates about the potential scale of this behaviour:

> [I]f the appeal character is such an important feature of the suicidal attempt as we have made it out to be, is there not a likelihood that this powerful and dangerous appeal will be used more and more, especially in a society which has made every individual's welfare its collective responsibility? I think that this danger can easily be over-estimated. 'Attempted suicide' is a behaviour pattern which is at the disposal of only a limited group of personalities.[1]

Two things deserve comment in this passage: the statement about society, welfare and collective responsibility, and also how Stengel is incorrect about the potential for the phenomenon to spread. We can see that Stengel is aware of a possible connection between the collective approaches to welfare and a socially embedded 'appeal'. He sees this in rather practical terms as potential to be exploited. We can see it slightly differently – as a connection between the political climate and a psychological object. A concern about social life, welfare and social work brings this object to light and constitutes it through the practical ministrations (interviews, home visits, follow-up and so on) detailed in the previous chapter. We shall return to this explicit connection of collective responsibility for welfare and this particular form of self-harm in the Conclusion – contrasting it with emergent neoliberal approaches that gain traction in the late 1970s.

The second point is that hindsight proves Stengel wrong, but sociologist Raymond Jack argues that criticism on this basis is unreasonable.[2] Stengel is not alone in this lack of foresight. A 1958 speech by Kenneth

Robinson, the most active Parliamentary agitator for suicide law reform shows how the problem of attempted suicide is small, even then. He claims that 'I am not suggesting that this is a vast problem, but our attitude to it in some ways symbolises what we think about human frailty and about mental illness'.[3] Rather than critique or excuse a lack of predictive power, this chapter asks a different kind of question: How is attempted suicide transformed from a behaviour pattern available only to a 'limited group of personalities' in the early-to-mid-1950s, to what one clinician calls 'a major epidemic' by the mid-1960s?[4]

This way of approaching the epidemic opens up a philosophical (ontological) question around what we mean by 'incidence'. When the recorded number of attempted suicides increases, what is happening? What is the relationship between the statistics and the real number of people performing this action? This question is unanswerable, and I do not think that it is particularly useful to conceive of these issues in this way. It is more useful to analyse how the numbers come to increase, how people become more aware of the problem, and how institutions become more adept at recording these ambiguous attendances at hospital.

The increased availability of mixed psychological and somatic scrutiny allows ambiguous intent to be projected – in a consistent and routine way – into incidents of self-harm presenting at hospitals. The epidemic remains fundamentally constituted by the practices through which it is recorded and administered. This chapter shows how the integration promoted by the Mental Health Act (1959) and the opportunities for government regulation presented by the Suicide Act (1961) combine to lay the foundations for epidemic self-harm in Britain.

By removing all legal obstacles to the treatment of mental illness in general hospitals, the legal changes contained in the Mental Health Act (1959) enable the further integration of mental and general medical therapeutics. Even the separateness of the observation ward is considered undesirable by some after 1959. The Suicide Act (1961) decriminalises attempted suicide, which had only arbitrarily been considered a police matter even in the 1920s, and even more rarely after the inauguration of the NHS in 1948 (this is the sense in which the problem is 'not vast' for Robinson). However, the law change means that the government finally feels able to act in a prescriptive way, intervening in the management of attempted suicide and actively promoting psychiatric attention, something that is much more difficult when the act is technically a common-law misdemeanour.

Government intervention aims to make referral to a psychiatrist from A&E consistent on a nationwide scale. This multiplies the possibilities for an epidemic (although without providing any extra resources). Attempted suicide as communication thus becomes a coherent national concern, but the resources available are insufficient to project a consistent social constellation around the physical injury. However, this basic coherence means that wherever appropriate resources are provided, the object can be found in abundance: an epidemic.

The Mental Health Act (1959): psychiatry into the 'mainstream of medicine'

Self-conscious efforts to achieve the equivalence of mental and physical medicine reach their zenith during this period, but have a broad history and continuing contemporary relevance under the banner 'parity of esteem' between mental and physical healthcare.[5] Whilst in one sense these concerns span the twentieth century and before, they remain contextually specific. Integrative efforts in the 1950s and 1960s based around psychiatric provision at general hospitals deserve special consideration; they are exceptionally self-conscious attempts at integration. The observation ward remains important in this process: many wards become treatment units in line with the prescient 1930s views analysed in Chapter 1 (as well as psychiatric liaison and referral services becoming more established). More broadly, the slowly changing functions of observation wards (see the previous chapter) play a key role in a re-articulation of attempted suicide. All these moves towards increased psychiatric provision enable the transformation of a physical injury arriving at a hospital into an interpersonal disturbance.

Two narratives: the dominance of 'asylum-community' and economic concerns

The historiography of the Mental Health Act (1959) significantly underplays its role in these integrative efforts. At the time, Kenneth Robinson draws out two distinct threads, noting that although the Percy Commission's Report and subsequent 1959 act are complicated, two more or less simple threads run through both: first, 'all distinction, legal, administrative and social, between mental and physical illness should as far as possible be eliminated'. Second, people who do not require long-term inpatient care should 'receive care and treatment while remaining in the community'.[6] It is this second thread that dominates the historiography of mental health in the twentieth century – the move from 'asylum to community'. The report and the 1959 act are conventionally and broadly seen as marking a shift from 'institutional' or 'asylum' to 'community care' (termed deinstitutionalisation or decarceration).[7] This narrative also centrally acknowledges that 'the aspirations of the Percy Commission were never fully supported in legislation since…no additional money was made available'.[8] The mobilisation of political concerns around this idea of a gap between the idealism of the report, and the financial provision for community care is one reason why the institution–community binary remains durable.[9]

This focus, oscillating between institutions and the community, sits uneasily with this account of attempted suicide as it neglects general hospitals and observation wards. Rogers and Pilgrim retain the emphases of asylum and community even when discussing general hospitals. Their assessment of District General Hospital (DGH) psychiatric units is that 'asylum theory and practice [are transposed into] DGH units and no new evidence of staff involvement with the communities of the patients they admitted'. Notions of asylum theory and a neglected community structure the analysis. Even more strikingly they characterise the Royal Commission on Lunacy and Mental Disorder of 1924–26, as containing an 'emphasis in 1926 on outpatients' clinics and observation beds in general hospitals (i.e., not in asylums)'. Their clarification of the significance of 'beds in general hospitals'– not in asylums– is revealing of their focus, between asylum and community: general hospitals are significant because they are not asylums and are bundled in with outpatient clinics.[10] Instead of making the DGH part of an asylum–community narrative, the present approach draws from Nikolas Rose's argument that 'rather than seeking to explain a process of de-institutionalisation, we need to account for the proliferation of sites for the practice of psychiatry'.[11] Different sites mean different contexts that require and sustain different kinds of practice. Focus on the DGH is an important part of the answer to Eghigian's question: '[W]here is psychiatry taking place?'[12] The clinical object, 'attempted suicide', emerges at the interface of psychiatric and general medical fields, and this is reconstituted by the 1959 act. Thus, much of the specific mental health policy discussion is not immediately relevant.[13]

The standard (somewhat neglected) narrative of integration, described in the Introduction, runs almost seamlessly from the Mental Treatment Act (1930), through the NHS (1948) to the Mental Health Act (1959). Charles Webster casts the 1959 act as tying up the loose ends left by the NHS in the march towards (presumably) fully integrated, comprehensive healthcare. He argues that 'the major loose end that was left by the NHS was the law relating to lunacy, and this was duly undertaken in

1959, following the Royal Commission on the Law relating to Mental Illness and Mental Deficiency'.[14]

In 1957 this commission (the Percy Commission) publishes its report, which contains the clearest and most widely circulated statement that psychiatry should become integrated with general medicine: 'Disorders of the mind are illnesses which need medical treatment...most people are coming to regard mental illness and disability in much the same way as physical illness and disability'.[15] It is stated in the text of the Mental Health Act, 1959, that '[n]othing in this act shall be construed as preventing a patient who requires treatment for mental disorder from being admitted to any hospital'.[16] Barbara Wootton demonstrates the sheer number of groups that are rhetorically committed to the integration of mental and physical medicine during the 1950s, citing evidence submitted to the Percy Commission. This includes testimony on behalf of the Association of Municipal Corporations ('it is now agreed that mental illness is a medical condition requiring the same amount of care as any other medical condition'); and the Royal College of Physicians ('the procedure for treatment of the mentally ill should approximate as far as possible to that of the physically ill'). The County Councils Association make 'suggestions for "accelerating" the "process of gradually placing the treatment of medical or physical illness on a similar footing"'; and the Association of Psychiatric Social Workers takes it as read that to bring 'the treatment of nervous and mental disorders more closely in line with that of physical illness' is a positive step. Wootton is clearly justified in stating that '[t]he wish to assimilate the treatment of mental and physical illness is thus widely supported'.[17]

This assimilation is broadly attempted by providing for the treatment of mental disorders in the same places as physical disorders – general hospitals. The increase in attempted suicide as communicative self-harm is founded in general hospitals. The vast majority of the time, it is the uncontroversially physical aspect of attempted suicide that first brings it to medical attention. Even when arguing in 1963 that all attempted suicides should be investigated by a psychiatrist, David Stafford-Clark remarks that it 'has surely never been suggested' that 'general physicians were to be wholly excluded from the management of these cases'.[18] Neil Kessel notes in 1965 that 'it is as a general medical problem that the poisoned patient first presents'.[19] This management, be it surgical or toxicological, is not performed in – nor is particularly relevant to ideas of – the community; it is vital not to conflate processes of integration with those of decarceration or community care. The emergence of a psychiatrically inflected attempted suicide in the second half of the twentieth century in Britain does include a sense of 'community'– the psychosocial setting – but one that cuts across canonical mental-health histories.

Ad hoc referrals and eclectic clinicians

This wider rhetoric of integration informs a number of idiosyncratic and ad hoc practices that bridge the separate regimes of general and mental medicine. A number of studies of attempted suicide are carried out at general hospitals in the late 1950s but not in observation wards. Therapeutic regimes are negotiated in various ways, turning physical injuries into psychosocial disturbances to varying degrees. Whilst the late 1950s and early 1960s seem to represent the rhetorical height of integration, the picture is much messier in terms of practical arrangements and clinical objects. What remains key is the intellectual, practical, interpretive labour that inscribes this 'attempted suicide' into casualty records, undercuts the significance of somatic injuries and constructs psychosocial environments around the attempts.

Studies of attempted suicide issue from a group of casualty departments in Gateshead (1953–7), Guy's Hospital in London (1958) and Birmingham (1959).[20] These studies negotiate the institutional obstacles between mental and general medicine in hospitals by arranging referrals of casualty patients to psychiatrists, enabling socially directed explanations for self-harm to various degrees that they term attempted suicide. The most colourful (and seemingly commonsense) analyses emerge in a study by John Lennard-Jones and Richard Asher from Central Middlesex Hospital (1959). They coin the term 'pseudocide' for these actions. The following illustrations show how quite socially embedded these attempts are, and how much questioning is necessary to situate them in this way. Under '[d]oubtful suicide attempts' they set out a detailed case-study description of a social situation, both before and after the 'attempt':

> A Hungarian girl, aged twenty, took 15 aspirins because she felt lonely when her Irish boy friend did not visit her at the weekend, and had been offhand when she telephoned him. She took the aspirins impulsively and was glad when she came to no harm. Next day a solicitous boy friend escorted a smiling girl from hospital. *Comment*: Suicide may have entered her mind, but the appeal value of her action was enormous.[21]

Under 'Spurious Suicide Attempts', they bring preceding and subsequent social situations to relevance again:

An Irish maid of twenty, working in a hotel, gave in her notice and was due to leave the next day. Having no friends in England and only a week's wages she felt that desperate action was needed. She swallowed a bottle of aspirins and then, having told the manageress what she had done, she undressed and went to bed. The doctor, urgently summoned, found her sitting up in bed combing her hair, but as he entered the room she fell back groaning ... *Comment*: A silly girl who liked showing off.[22]

These descriptions are folksy and idiosyncratic, but draw upon Asher's well-established interest in psychology. The intent in these cases is articulated through common-sense ideas of communication: 'appeal value' and 'showing off'. Despite the casual tone, the practices used to elicit these objects are remarkably labour-intensive. The information used to construct the above case histories is only fully obtained 'after carefully, and sometimes repeatedly, questioning patients and their relatives'.[23]

Thus at Guy's and the Central Middlesex in London, in Gateshead and in Birmingham, 'attempted suicide' emerges. Referral enables a series of transfers between separated therapeutic regimes. In Asher's case, it is his eclectic (boundary-crossing) interests that are crucial. The object appears with increasing frequency, and yet the irregular, impermanent nature of the practices negotiating the split makes these clinical objects seem like so many miscellaneous, disconnected occurrences. There is certainly not much sense from the articles surveyed that attempted suicide is a national problem. The potential for an epidemic is clearly there, but it requires more high-level coordination and intervention to be fully realised.

As the provision of mental-healthcare is rethought and reconstructed in the late 1950s, new objects appear. Too great a fixation on 1959 is unhelpful because the act removes restrictions to mental treatment. These are largely irrelevant, in one sense because this particular phenomenon presents first as physical injury. The 1959 act does not enact integration, it merely removes legal obstacles. Whilst the 1961 retraction of the law from suicide and attempted suicide is similar in one sense, the government is much more pro-active, prescriptive and practical, so the Suicide Act repays this kind of closer scrutiny.

Suicide Act 1961: complex intent, legal reform and government intervention

The decriminalisation of suicide and attempted suicide in 1961 decisively ends some longstanding medico-legal debates around suicide.

These debates are important, as several legal-reform arguments bring complex intent to prominence, and the resulting retraction of the law initiates a far-reaching shift, enabling an openness and formality around the treatment and recording of attempted suicide. After the act is passed, the Ministry of Health recommends, on a national scale, that all cases of 'attempted suicide' seen at casualty or by GPs are considered for referral for psychiatric assessment. This positive intervention thus multiplies the possibilities for the (re)articulation of this phenomenon. Rates of psychiatric referral of 'attempted suicide' are actively followed up, policed and collated by the Ministry of Health; hospital groups have to account for any significant number of patients not directed to psychiatric scrutiny. The rhetoric around 1959 encourages integration, but these developments prescribe crossover, fuelling the growth of this phenomenon from a 'limited number' to an 'epidemic'.

The Suicide Act as a tale of two conflicts

The Suicide Act of 1961 has yet to receive sustained attention from historians. It is instead viewed as a minor part of the clutch of legislative changes and government reports seen to constitute the first 'permissive moment' in post-war Britain, under the reforming Conservative home secretary, Richard Austen Butler, between 1957 and 1962. (The second of these is related to Roy Jenkins's time at the Home Office, 1965–7.) Butler's time as home secretary sees discussions around 'how far to liberalise social constraints (if at all), particularly in relation to gambling, licensing, Sunday observance, suicide, censorship and the law governing sexual behaviour'.[24] These discussions play out against the intellectual backdrop of the most famous jurisprudential debate of the twentieth century, between Lord Patrick Devlin and Professor Herbert Hart.

The debate is sparked by the 1957 publication of the Wolfenden Report, which recommends (among other things) that 'homosexual acts' be decriminalised between consenting adults in private.[25] This debate snowballs into something much more general: in Peter Hennessy's apt summary, 'at issue was the power of the state to outlaw private practices it deemed immoral even if they harmed no one else'.[26] Devlin, a judge and later a Law Lord, argues that the law must be involved with moral questions because there can be no theoretical limit to society's powers to police itself. He argues that 'the criminal law could not operate without a moral law'.[27] Hart, a philosopher and professor of jurisprudence at the University of Oxford, counters that moral questions are outside the legitimate remit of the criminal law, unless they involve harm to another person (following such nineteenth-century liberal philosophers as John Stuart Mill). The Suicide Act of 1961 features explicitly in this debate, as Hart praises the decriminalisation of suicide as 'the first Act of Parliament for nearly a century to remove altogether the penalties of the criminal law from a practice both clearly condemned by conventional Christian morality and punishable by law'.[28]

Mark Jarvis's study of the reforming Conservative government of the late 1950s and early 1960s is subtle and discerning, but rather rushes through the reform of the law relating to suicide, allotting it fewer than three pages. The act figures most prominently for Jarvis as a site of personal/political tension, an opportunity for the expression of the differing political dispositions of Butler and Prime Minister Harold Macmillan. Although the act is strictly out of the time period of Hennessy's *Having it So Good: Britain in the Fifties*, he uses the act in a very similar way. Both analyses pivot around an exchange between Macmillan and Butler. Macmillan asks: 'Must we really proceed with the Suicides [*sic*] Bill? I think we are opening ourselves to chaff if, after ten years of Tory Government, all we can do is to produce a bill allowing people to commit suicide'.[29] Butler counters: 'The main object of the Bill is not to allow people to commit suicide with impunity ... It is to relieve people who unsuccessfully attempt suicide from being liable to criminal proceedings'.[30]

For Jarvis, this emphasises 'a wider sense of tension between the Home Secretary and Prime Minister ... In his flippant attitude to reform of the suicide law, the Prime Minister showed how detached he had become from social reform, and antagonised Butler with his lack of insight at a time of major change'.[31] Hennessy prefaces the exchange with the contention that 'Macmillan's detachment, verging on insouciance, really irritated Butler'.[32] Both accounts go beyond the accessible and human narrative around personalities to make both this exchange and the act function as sites for the Hart–Devlin debate. Jarvis argues that 'in the case of the law governing suicide, Butler had modernised regulation by shifting it from a religious basis towards a more clearly defined border between law and private morality'.[33] For Hennessy, this exchange shows that 'Butler was, by nature and intellect, in the Hart camp'.[34]

Suicide law reform is thus placed firmly in the context of the Hart–Devlin debate, as a jurisprudential and parliamentary expression of moral libertarianism. This obscures much of its complicated resonance. Instead of positioning it within a programme of liberal reforms, or as a barometer of political instincts (liberal utilitarianism versus moral paternalism) lurking beneath political rivalries (reformist home secretary versus traditionalist prime minister), or even as an expression of a celebrated jurisprudential debate, the analysis here shows how the act initiates changes in hospital practices, setting in train processes that enable, constitute and sustain a specific epidemic of self-harm as communicative overdose. It is important to draw a distinction between the retreat of the criminal law from concerns articulated in moral and psychological language, and the much larger retreat of the state from social management and support in the 1980s. This 1950s legal reform is carried out in the context of a sustained commitment to social and psychological support – as will become clear.

Stengel, legal reform and complex intent

The roots of the 1961 act can be most clearly seen – purely in parliamentary terms – in the repeated questions of Kenneth Robinson, Labour MP for St Pancras North, whose richly varied reforming political career involves: being the first chairman of the National Association for Mental Health; minister for health in the Labour government of 1964–8; sponsor of a Private Member's Bill to legalise abortion in 1961; and member of the Homosexual Law Reform Society's executive committee. Robinson begins asking questions of Butler on 6 February 1958. Butler's initial response is that he is 'not satisfied that any change in the law is desirable'. When Robinson counters that 'considerable and growing opinion in the medical and legal professions, and among the general public' is in favour of a change, Butler neatly refocuses the issue away from medical and legal professionals, and onto what he imagines to be much safer ground: 'the present concept of suicide as a crime has its roots in religious belief'.[35]

Robinson's reference to 'growing opinion' denotes a late-1950s surge in debates around the law on suicide. This includes Glanville Williams's *The Sanctity of Life and The Criminal Law* (1958), the British Medical Association and Magistrates' Association Committee's (BMA-MA) second report (1958) in just over a decade (having also produced a memorandum on suicide law in 1947) alongside a contribution from the Anglican Church, *Ought Suicide to be a Crime?* (1959). A brief look at these and other texts shows that as well as being explicitly influenced by Stengel's work, legal arguments in favour of reform promote visions of complex and ambiguous intent driving 'suicidal' actions.

Against this model, perhaps the earliest post-war contribution in favour of decriminalisation – that the sanction of the law is no deterrent because that person concerned expects to be dead – implies an attempted suicide modelled upon straightforward, genuine intent. The

British Medical Association's 1947 memorandum, prepared by their Committee on Psychiatry and the Law, explicitly downplays the significance of communicative or so-called hysterical attempts:

> Whether the prospect of police court proceedings is in any way a deterrent to the would-be suicide is a question which may be asked. Except in respect of hysterics whose motive, though they may not be aware of it, might be to attract attention, the large majority of those who attempt suicide do so in the expectation of completing the act. Thus it is probably true to say that would-be suicides are not likely to be deterred by fears of police court proceedings, since they believe they will be dead before the issue arises.[36]

The power and significance of the deterrence argument in this case is connected to debates circulating at that time about the non-deterrent effect of the law on capital punishment. Although hysterical attempts are downplayed in the context of these arguments about decriminalisation, there is still an acknowledgement that suicidal intent can be complicated.

Glanville Williams, eminent legal scholar and conscientious objector to the Second World War, publishes his controversial *The Sanctity of Life and The Criminal Law* in 1958. The book ranges widely, examining the philosophy behind prohibitions of contraception, sterilisation, artificial insemination, abortion, suicide and euthanasia. His arguments for the decriminalisation of suicide are noted by the Home Office and in Parliament, adding considerable intellectual muscle to reform arguments. His position shows how the concept of communicative attempted suicide can complicate (and critique) the law in a new way. The idea of self-harm as communication gains traction in the law-reform movement because it is used to undermine the law by scrutinising suicidal intent. Williams argues that '[m]uch light has been shed upon [attempted suicide]...by a recent medical study made by Professor E. Stengel and Miss Nancy Cook'. He also draws upon Lindsay Neustatter's *Psychological Disorder and Crime* (1953). One of Neustatter's examples in which the police will take action and prefer criminal charges is when 'repeated attempts have been made, and it is evident that these are not genuine, but due to sensation-mongering: e.g. a girl several times threw herself down into shallow water where she could not possibly drown'. Williams's keen legalistic analysis brings out a tension in the law's operation: 'If an attempt is not seriously intended, it is not, in law, an attempt, and neither a prosecution nor a conviction is justified. There is no crime of attempted self-manslaughter by knowingly running the risk of death'.[37]

Part of Williams's critique of operation of the criminal law is thus based upon his reading of Stengel and Cook. He argues that under the umbrella of suicidal acts there are three important sub-categories: the genuine, the demonstrative, and between those lies the gamble, which Williams claims is 'consciously an attempt at suicide, but unconsciously a gesture':

> The three kinds of suicidal acts call for separate consideration from a legal point of view. Genuine attempts at suicide are offences under present English law. Suicidal demonstrations are not, as such, offences. The legal status of the third group is undetermined; indeed, no court has yet had to pronounce upon unconscious motivation in criminal law. It seems probable, however, that such motivations, even if proved to the satisfaction of the court, will be ignored, on the ground that legal sanctions can only deal with the conscious mind.[38]

Whilst only one of the three categories is conclusively deemed ineligible through Williams's mobilisation of Stengel and Cook, the ambiguously motivated attempted suicide popularised by them has specific traction in the reform arguments. In Williams's hands it involves a statement that the law as it stands is not relevant to a gestural kind of attempted suicide.

Geoffrey Fisher, the Archbishop of Canterbury, forms a Church of England committee chaired by J.T. Christie, his direct successor as headmaster of Repton public school. In 1959 this committee issues the booklet *Ought Suicide to be a Crime?* A key member of the committee is Doris Odlum who, as a psychiatrist and magistrate (and later a president of The Samaritans), also sits on the joint BMA-Magistrates' Association committee. The booklet is written in three parts, with distinctly legal, psychological and religious arguments marshalled in turn.

In the legal section there is the argument that undercuts the law's application, as in Williams' and Neustatter's analyses: 'The man who repeatedly throws himself under a 'bus is plainly a public menace, but there cannot be many such men. It is doubtful whether, as a matter of law, anyone can be properly convicted of attempted suicide unless it is proved that he or she intended to kill themselves'. Again, the law is seen to be of ambiguous relevance when intent is scrutinised. Even the section that approaches the question from an explicitly moral and religious angle invokes an elastic notion of a 'complex mental history' to question the idea of intent: 'Much more is now known about suicidal tendencies and about the complex mental history that can mobilize a potential suicide. It would seem as if there are not many suicides which can nowadays be regarded as wholly voluntary and deliberate'. Psychiatric advances are mobilised to question whether a legal response is appropriate: 'As a result of the development of psychiatry, it can be granted on all sides that many cases of suicide and attempted suicide should never be legally assessed at all, nor religiously condemned'.[39]

This 'development of psychiatry' is most likely a reference to the removal of legal formalities in the 1959 Mental Health Act. As a July 1959 speech on this bill in the House of Lords shows, the issues of suicide and mental-health law reform are connected, as 'one of the commonest kind of mental patients coming before the court ... [is] the attempted and unsuccessful suicide'.[40] The Mental Health Act is concerned with the relationship between legal sanction and psychiatric treatment. This brings attempted suicide to prominence because that action is considered a psychiatric problem and is also against the law. Thus law-reform arguments can bring to new prominence complicated or ambiguous intent around suicidal actions.

Returning to passage of the bill, on 6 March 1958 Robinson informs Parliament that he has obtained over 170 signatures to a motion for reform of the suicide law. He argues pointedly that his motion had been signed by those 'of all shades of religious opinion'. Butler again attempts to deflect rather than deal with the issue directly, suggesting that '[i]f the Opposition would wish to find time on a Supply Day for this or any other similar general question, it would be an interesting subject for the House to discuss'. Undeterred, Robinson submits a question a week later, asking 'on what evidence he bases the view that amending legislation to remove suicide and attempted suicide from the list of criminal offences would not be generally acceptable to public opinion'. Rather testily, Butler's reply is that '[e]xperience suggests that changes in the law on matters which involve religious and moral issues are likely to be contentious'. However, he is publicly more open about the possibility for legislative change, adding that 'I have not closed my mind on this Question and am continuing to study it carefully and sympathetically'.[41]

At the end of May, Robinson applies more pressure, mentioning the memorandum issued by the Joint BMA-MA Committee; in October, he criticises the law on the grounds that it is no deterrent: 'Clearly, the fact that suicide or attempted suicide is an offence against the law has very little, if any, effect on the mind of the would-be suicide'.[42] Butler directs the Criminal Law Revision committee to look into the practical aspects of changing the law in 1959, and Robinson keeps up the pressure, eight times posing oral and written questions about the progress of the committee. The bill is introduced in the Lords on 14 February 1961 and is finally enacted on 3 August that year.

Hospital Memorandum HM(61)94 – Prescribing referral between therapeutic regimes

After attempted suicide is officially decriminalised in August 1961, in September the Ministry of Health issues Hospital Memorandum HM(61)94 'Attempted Suicide'. It asks 'hospital authorities to see that all cases of attempted suicide which come to their notice receive adequate psychiatric care'.[43] Attempted suicide is again inextricably bound up with negotiation between separate therapeutic regimes – from the acute, somatic medicine of casualty departments to psychiatric care. However, no extra resources are provided to casualty departments to enable this referral. In any case, similar to the previous chapter's analysis of A&E, the intensive scrutiny required for this object to flourish remains ill-suited to the administrative co-ordinating that occurs in 1960s casualty departments. Simply providing for referral or crossover is insufficient to sustain a psychosocial attempted suicide. However, it does attempt to coordinate referral on a nationwide (potentially epidemic) scale.

The idea behind HM(61)94 is first mentioned in correspondence between the Home Office and Ministry of Health on the final day of 1958. The latter department assumes responsibility for the promotion of psychiatric referral in cases of attempted suicide. Civil servants consult widely in mainland Europe and North America, asking their health department counterparts how such cases are dealt with under various legal arrangements. At a subsequent meeting between representatives from the Home Office, Health Ministry, British Medical Association and Magistrates Association. It is noted that

> in a great many cases the person would have been admitted to hospital to receive treatment for his physical injuries. At present, however, many of these persons were discharged without a psychiatric examination. The nature of the offence suggested that such an examination would be advisable in all cases … this was a matter on which the Minister would be prepared to give guidance to hospitals.[44]

The purpose of the memorandum is to ensure that the physically injured attempted-suicide patient obtains psychiatric assessment at general hospitals. Government intervention is needed to integrate the two therapeutic regimes that formally and legally become equal after the Mental Health Act 1959.

This cause receives extra impetus in November 1960 when the Royal Medico-Psychological Association (RMPA) produces a report on Casualty and Accident Services, written by W. Linford Rees and John S. Stead. At this point, Rees is chairman of the Research and Clinical Section of the RMPA, having spent formative War years at Mill Hill conducting research at the Effort Syndrome Unit, the start of his work on psychosomatic disorders. He is remembered as facilitating 'the work of psychiatrists within the context of the general hospital'.[45] This document is part of a more general early 1960s concern about casualty departments, which leads to the publication of a number of critical and anxious reports.[46]

Rees and Stead are critical about the general level of psychiatric care: 'In only thirteen of the forty nine hospitals was the casualty officer able to call in a psychiatrist to advise on disposal'. More disturbingly: 'Few of the hospitals in the regions and few of the London teaching hospitals felt that they had adequate psychiatric advice available for assessment and appropriate disposal of patients'. All the recommendations concern the integration of psychiatric and general medical expertise in general hospitals, covering the provision of initial advice, facilities for short-term psychiatric-diagnostic observation, and arrangements to transfer patients to either a psychiatric unit or psychiatric hospital.[47] Concerns about the practicalities of integration – specifically the number of consultant psychiatrists – are also present in the 1958 British Medical Association and Magistrates' Association Report on attempted suicide.[48]

Wider integration and legal opportunity: common ground between 1959 and 1961

Arrangements for the hospital memorandum on attempted suicide are taken in hand later in November 1960, primarily because '[n]ow that the government have announced their intention of amending the [suicide] law … the time has come for us [Ministry of Health] to issue a hospital memorandum urging hospital authorities to see that all cases of attempted suicide which reach them are given a psychiatric investigation'.[49] Senior civil servant Patrick Benner adds that '[i]t seems all the more necessary to go ahead with this fairly soon in view of the recent report of the Royal Medico Psychological Association suggesting that this is a matter on which a good many hospitals are not doing very well'.[50]

The two broad reasons – the opportunity provided by a government-sponsored bill to change the suicide law, and an appreciation that psychiatric advice in casualty departments is not all it should be – show up consistently in the memorandum negotiations and revisions. Instead of seeing the HM crudely, as solely enabled by the Suicide Act, it is significantly concerned with the wider integration promoted by the Mental Health Act 1959.

The Suicide Act might be a convenient prompt, but Benner argues that '[t]he general points we need to make to [hospitals] are valid even in advance of the legislation [because] our aim is to produce, in advance, the requisite degree of medical and social care'.[51] It is also claimed that there is 'good reason to think that hospital practice is in need of improvement now and this depends in no way on the outcome of the [Suicide] Bill'.[52] Whilst part of MH(61)94 is prompted by the legal change, integration of therapeutic regimes ('improving hospital practice') is a significantly wider issue. This is the shared territory between the 1959 and 1961 acts.

Stengel takes a narrow legalistic line, rather than credit the government with any serious acknowledgement that psychiatric facilities are inadequate in A&E. He writes:

> The role of the psychiatrist in the management of attempted suicide in the general hospital has for the first time been officially defined. Apparently, once the problem of suicide was taken out of the hand of the law, the Ministry of Health considered that the health authorities had to accept responsibility and to advise how it should be discharged.[53]

The transformations in the previous chapter at observation wards are here promoted at accident and emergency departments. After 1961, the possibilities for the emergence of communicative attempted suicide are transformed in size and scope, the foundation for a problem of epidemic proportions and national significance.

The text of the memorandum is centrally concerned with integrating psychological scrutiny into the overwhelmingly somatic focus of casualty departments. It is stated that '[t]hese cases often come to hospital casualty departments for urgent lifesaving physical treatment...after physical treatment the patient is sometimes discharged without any psychiatric investigation of his condition [which is] of major importance in most cases of attempted suicide'. It continues, offering suggestions heavily influenced by Rees's and Stead's report: 'Hospital authorities are therefore asked to do their best to see that all cases of attempted suicide brought to hospital receive psychiatric investigation before discharge … Where the hospital has no psychiatric unit, it may be necessary to arrange for liaison with a neighbouring psychiatric hospital'.[54] Again, arrangements negotiating the split between psychiatry and general medicine are necessary for this clinical object to thrive.

Stengel does not see the potential for attempted suicide to multiply exponentially as a result of the Suicide Act, rather curiously focusing instead on coroners and completed suicide figures: 'Psychiatrists do not expect the law to lead to an increase in suicidal acts, but a slight rise in the suicide figures will not be surprising … some coroners may be less hesitant about giving a verdict of suicide rather than an open verdict'. He does make a concession:

> It is also possible that the number of attempted suicides diagnosed as such in the hospitals may show a slight increase. If so, this should not be taken at its face value … Some hospital doctors were known to refrain occasionally from referring to the suicidal attempt in their diagnostic formulations, in case their patients should suffer inconvenience. For the same reason, the protestations of some patients that they had taken overdoses of dangerous drugs without suicidal intention may have been accepted too readily. Small increases in the numbers of suicides and attempted suicides in the next few years can therefore be regarded as artefacts.[55]

Unsurprisingly, Stengel remains within traditional ideas of incidence, seeing institutional change as effecting variations upon a real total number. Instead, the argument pursued here is that changes in organisation are fundamental to the kinds of numbers that are produced. The Ministry of Health also does not see this as a problem on a huge scale, as when the hospital memorandum is finally issued, it is decided not to alert the press because '[t]he documents are self-explanatory, and the subject, though important, is of limited scope'.[56] With hindsight, the foundations are there, but traditional ideas of incidence obscure the epidemic potential from even the most vocal publicist for attempted suicide.

Psychiatric resources and ministry follow-up

The A&E department is the site at which the Ministry of Health seeks to intervene, to entrench referral practices between general medicine and psychiatry. However, there are no extra resources provided for the proposed extension of psychiatric referral. Stengel optimistically believes that HM(61)94 will be a stimulus for the establishment of psychiatric outpatient departments and DGH psychiatric units, and for social and community services in general:

> Considering the large number of consultations required by the Ministry of Health [Hospital Memorandum]...The pressure for additional psychiatric staff and for the creation of more psychiatric outpatient departments is likely to increase. This will be all to the good because it will make the community aware of the inadequacy of the psychiatric services and will speed up plans for creating psychiatric departments in general hospitals. Thus, attempted suicide, that last and supreme appeal for help, may act as a powerful stimulus for the improvement of psychiatric and social services.[57]

This again shows the link between the two acts of Parliament analysed in this chapter. However, the idea that a newly decriminalised attempted suicide might stimulate the integration of mental and general medicine is rather back-to-front. The much broader efforts attempting the integration of therapeutic regimes are what enable this object to be constituted – that are fundamental for the emergence of this supreme appeal for help. S.W. Hardwick of the Royal Free Hospital writes to the ministry and makes the same point as Stengel, that there are insufficient resources to carry out all these referrals: 'If I am right in my interpretation of the H.M., a considerable amount of additional work and responsibility will have to be undertaken by the Psychiatric Department, which may mean a requisition for extra staff'.[58] The government's approach to integrating general and mental health in this specific case seems consistent with the broader (lack of) financial provisions around the Mental Health Act 1959. Stengel hopes that

> doctors and hospital authorities who have found the Ministry's recommendations impracticable will say so in no uncertain terms. It would be against the interests of patients to adjust the attempted suicide figures to the psychiatric resources available instead of adjusting the resources to the real demands.[59]

Given the importance that is placed throughout this book upon the high intensity of scrutiny necessary for this psychosocial self-harm to emerge consistently, casualty again seems like an unlikely candidate.

It is possible to glimpse the level of impact that the HM has on casualty services, because the Ministry of Health decides to follow up the recommendations. Benner sends a note to government statistician G.C. Tooth stating that whilst 'it is not our practice to follow up all H.Ms by any means...this is a rather important subject where I think some kind of action from us would be reasonable'.[60] Benner expands upon the importance of this statistical enquiry. He appreciates that the whole problem of attempted suicide has been passed to the health service and therefore 'it seems right that we should know how they are dealing with it'. Tooth agrees, emphasising that it is important for the ministry to have a sense of how many patients have been seen by psychiatrists before they are discharged from hospital – having had first aid for their injuries.[61] Integration of psychiatric and general medicine for scrutiny of patients arriving at casualty is not simply prescribed, but actively policed after the change in the law. Regional hospital boards are asked to submit the number of attempted suicides seen by a psychiatrist in the twelve months since the issue of the hospital memo. They are asked for the approximate number of cases, the proportion seen by a psychiatrist and details of any measures to improve rates of follow-up.[62] This is a concerted effort to prompt and shape casualty department practice. This information is collated and written up in an internal document in January 1964.[63]

The Ministry expresses broad satisfaction because although 'replies from Boards vary considerably...most managed to report that 75% of admissions were seen by psychiatrists'.[64] The memorandum prompts a number of diverse practical changes in various hospitals concerning psychiatric liaison. These are glossed illustratively here to give a flavour of the different ways in which the therapeutic divide is constituted and negotiated in the same move. The Sheffield Regional Hospital Board (RHB) report that the Sheffield No.1 Hospital Management Committee (HMC) has the lowest rate of referral to a psychiatrist in that region (65%). The hospital psychiatrist 'suggests a special form for all patients admitted for attempted suicide' as a remedy.[65] Grimsby HMC, under the same board, reports that '[s]ince HM(61)94 a rota of Mental Welfare Officers has been arranged whereby one sits in at each clinic and follow-up and all cases are referred to Consultant Psychiatrist in the Group'.[66] Under the North-West Metropolitan RHB, the Luton and Dunstable Hospital reports: 'During the last year the number of days on which there is a psychiatric out-patient clinic has increased from 2 to 3 a week, so that psychiatrist are more readily available to see these patients'. Under the same RHB, Mount Vernon Hospital achieves only 35% referral, and the psychiatrist concerned comments that 'unless he is relieved of some other commitments he will not have time to see all of the cases that should properly be referred to him'.[67]

This board claims in its covering letter that '[w]here the information ... shows a markedly inadequate service ... the possibility of improvement [will] be discussed with the members of staff concerned'.[68] For the Wessex RHB, '[t]he Board has taken action to bring the Salisbury Hospital group with a 39% return into line' and although the Isle of Wight reports that only 50% of cases have been psychiatrically assessed over the past year, '[i]n future all such cases will be seen by a Psychiatrist'.[69] These are uneven, ad hoc, idiosyncratic practices, despite the best efforts of the Ministry of Health. Referral arrangements involving mental welfare officers and psychiatric out-patient clinics exist alongside new memoranda, renewed efforts at referral to psychiatric consultants and mental hospitals that, despite their differences, are all attempting to move towards integration.

However, not everything goes so smoothly – Cardiff RHB even interprets the guidance in such a way as to decrease the visibility of communicative attempted suicide.[70] Stengel has other problems with it and implies that the return is less than useless. His letter to the Ministry of Health is unfortunately no longer in the file, but there remains a copy of one he sends to the Superintendent of the Royal Infirmary, Sheffield. In it he argues that 'I have not been able to comply with your request ... patients who have made suicidal attempts are not usually diagnosed as "attempted suicide" but under some other heading ... The only way to provide the required information would be for the Ministry to request hospitals to put "attempted suicide" into the diagnostic index'.[71] He says that 'it would be a pity if the Ministry should accept information which cannot possibly be valid [and] dangerously misleading'.[72] This is a significant problem for the emergence of a consistent, epidemiological object of attempted suicide.

As Stengel's criticism highlights, without either a customised structure for its record, or the labour-intensive scrutiny of research psychiatry, attempted suicide is exceptionally difficult to pin down. Specialised research projects begin to record it during the early 1960s. W. Malcolm Millar, George Innes and Geoffrey Sharp design a research questionnaire in the early 1960s that includes the question: 'Has a suicidal attempt formed any part of the present illness? Yes/No'.[73] Peter Sainsbury and Jacqueline Grad prepare a clinical record sheet for psychiatrists to record reasons for deciding upon a certain disposal option. Next to 'previous mental illness' there appears the phrase '(N.B., Suicide Attempt)'.[74] This reminds psychiatrists that a suicide attempt is to be considered as part of a mental illness (even perhaps a trivial one, apt to be dismissed as a gesture). However, recording attempted suicide here requires a special record sheet or specialised psychiatric research project. It becomes clearer why the ministry-backed crossover is insufficient on its own.

Finally, Medical Officer John Brothwood proposes to the Ministry of Health a statistical study of attempted suicide at A&E. It involves distributing a form to casualty departments in order to ascertain the methods and motivations behind attempted suicides. Several objections are raised about the definition of attempted suicide (by Eileen Brooke). Equally damaging questions about the practicability of obtaining the information are raised by a Dr. Otley: 'Many of the questions ... would be unanswerable or answerable on very scanty information "at the time of consultation"' by the medical officer in casualty. The scheme fails to gain approval because the casualty department is unsuited to the project, allowing only a small and inadequate amount of information to be collected. The complex definitional problems that circle around intent, which could enable the intent to become communicative, require those with background – inescapably social – knowledge.[75]

The limitations of casualty differ from those in some observation wards, where treatment and follow-up are more established. However, the casualty department and the HM that seeks to intervene upon it still attempt to negotiate the enduring boundary line between psychiatric and general medicine, and to draw out, control and produce information about attempted suicide. The inescapably social, communicative reading of attempted suicide needs more than just referral to and liaison with a psychiatrist. It needs consistent psychiatric scrutiny, and more of an institutional foundation and psychiatric resource base than a memorandum can provide. The efforts of MH(61)94 at securing nationwide rates of 75% referral do have an effect, prompting and solidifying channels of communication and scrutiny between accident departments and psychiatric expertise. However, a lack of extra resources and the sorting role of the casualty department within the NHS undercuts high-intensity psychiatric scrutiny at that site.

Concluding thoughts

There is a strong link between the Mental Health Act 1959 and the Suicide Act 1961. Both are implicated in a process through which different therapeutic regimes are integrated at general hospitals. This makes possible a consistent articulation of a highly psychologised, highly social reading of self-harm with complex intent. This contrasts starkly with today's clinical concern with self-cutting, which is based upon internal, and sometimes neurochemical triggers. Both acts of Parliament involve the removal or significant retraction of the law around the field of mental disorder (with suicidal behaviour securely, though not inevitably, entrenched as part of this field). This enables a more fluid interaction between mental and general medicine, altering the kinds of clinical objects likely to emerge. The Suicide Act, in removing the legal sanctions around attempted suicide does not necessarily change practices very much in one (empirical) sense; people are not being convicted very much during the 1950s. However, reform arguments have a resonant connection with ambiguous suicidal intent, and decriminalisation alters the terms of the debate through which attempted suicide is conceptualised, prompting formal intervention by the Ministry of Health.

Because of the high level of psychiatric scrutiny required to produce complex, communicative intent around a presenting physical injury, HM(61)91 does not enable a huge number of studies by itself. The lack of extra resources is significant, but perhaps even more significant is the vastly increased potential for the object to flourish in a number of different sites, if increased resources become available. This is another important step for the progress of a clinical object – from an observation ward curiosity to one inscribed in a nationally consistent manner. The epidemic – and the broad, homogenising administrative machinery required for a multi-site epidemic – emerges through wider integration promoted through a retraction of the law in the areas of suicide and mental health more broadly.

Returning to the notions of incidence broached discussing Stengel's attitude to the hospital memorandum, we can see that as the potential for this clinical object becomes more and more widespread and more visible, the behaviour potentially becomes more and more available. Ian Hacking observes:

> Cynics about one thing or another...say the epidemics are made by copycats. But even if there was a lot of copying, there is also a logical aspect to 'epidemics' of this type. In each case...new possibilities for action, actions under new descriptions, come into being or become current...to use one popular phrasing, a culturally sanctioned way of expressing distress.[76]

Hacking shows, in his example of multiple personality disorder, that this logic of epidemics is a powerful and useful way to understand how behaviours travel and multiply. Something similar happens with attempted suicide. His use of distress as a basic anchoring category also has a history. In the next chapter the growing resonance of terms such as stress and distress is analysed and placed into context. Psychological medicine increasingly turns to these concepts to understand mental disorder; attempted suicide is central in this development.

Except where otherwise noted, this work is licensed under a Creative Commons Attribution 3.0 Unported License. To view a copy of this license, visit http://creativecommons.org/licenses/by/3.0/

OPEN

4

Self-Harm as a Result of Domestic Distress

Minister of Health Enoch Powell's *Hospital Plan for England and Wales* (1962) is a familiar landmark in twentieth-century psychiatry.[1] In 1961 Powell's 'water tower' speech to the National Association of Mental Health eloquently launches the ideas contained within the plan.[2] It is an evocative portrayal of asylums as grand, obsolete monuments to Victorian ideas of mental-health care. There is much historiographical focus upon how the plan augurs the scaling back of mental inpatient provision, but much less on how it signals the broader uptake of a new model of integration between psychiatry and general medicine. This model, based upon the establishment of psychiatric units in district general hospitals (DGHs), involves a more intimate connection between general hospitals and psychiatry than do observation wards. The DGH psychiatric units promoted by the plan undercut the progressive status and bridging function of the observation ward.

A variety of referral practices, shown in the studies analysed below, demonstrate again how a certain kind of (socially directed) self-harm emerges according to the practices used to bridge the gap between separated therapeutic approaches of general and psychological medicine. However, whilst this does lead to an increasing number of studies producing a socially embedded 'attempted suicide', it also shows the limits to how far these approaches can converge upon patients. The approaches of general medicine (as well as specialisms such as surgery), are arranged and administered very differently to psychological medicine inside DGHs. These approaches remain persistently separate, and the new arrangements designed to bridge this gap and focus psychiatric scrutiny on physically injured patients provoke conflicts over resources.

This chapter shows how attempted suicide is developed beyond psychological scrutiny in A&E and mental observation wards. The first half focuses upon a Medical Research Council psychiatric research unit in Edinburgh, where this object is reframed as 'self-poisoning'. This narrowing of focus from all the potential methods of damaging oneself to just one passes curiously unremarked. The self-evidence of poisoning seems clear in the 1960s, even if it is not so now. A strong research base enables Neil Kessel, Norman Kreitman and others to unify these cases under the blanket term 'distress' and project their causes into domestic space. The combined facilities for psychiatric evaluation and resuscitation (as well as access to PSWs) available at Edinburgh's Ward 3 are not widespread. After 1965, a number of studies emerge from various places – largely focused upon psychiatric units in District General Hospitals (DGH). The second half of the chapter analyses how these different hospitals begin to focus upon communicative attempted suicide. Its growing stability, intellectual credibility and increasing public health profile mean that it becomes fully established as an epidemic phenomenon. In 1969 it is renamed 'parasuicide' by Kreitman and others.

This development of self-poisoning and attempted suicide continues to make sense as part of the broad turn to the social setting. The social setting's impact upon mental health and well-being is described through concepts of stress, distress and coping, as well as the practical ministrations of social work. These developments allow mental disorder to be further reconceptualised as an interpersonal, fundamentally social phenomenon. Again, the effects of a broad political commitment to state-sponsored social support is clear. On a practical level, psychiatric social workers remain key to communicative self-harm (especially at Edinburgh), using the practice of home visiting to root this object in a pathologised domestic environment. Once it is established in domestic social space, this space is then increasingly presumed to cause self-poisoning. This still relies upon interactions between therapeutic regimes, but brings increasingly gendered dynamics of domesticity and emotionality into play.

In January 1959 eminent psychiatrist Denis Hill gives a talk to the MRC assessing their psychiatric research policy. Having studied neurology before the war, Hill succeeds Aubrey Lewis in the chair at the Institute of Psychiatry in 1966. In this 1959 review, he suggests the establishment of two psychiatric research units, one in psychiatric genetics, under Eliot Slater, the other in psychiatric epidemiology under George Morrison Carstairs.[3] Carstairs's unit becomes a central site for the study of attempted suicide during the 1960s. Formally named the Unit for

Research on the Epidemiology of Psychiatric Illness, this research unit (especially the work of Assistant Director Neil Kessel) focuses a high level of psychological research resources upon attempted suicide, developing the potential provided by mixed therapeutics and establishing a number of stereotypical characteristics for those thought to communicate their distress through self-poisoning.

The start of the unit's life in the late 1950s is chaotic. It is initially sited in London at University College Hospital, but when Alexander Kennedy, professor of psychological medicine at Edinburgh dies, Carstairs is awarded the chair and takes the unit with him in April 1961. Carstairs becomes the honorary director, and his heavy clinical and teaching commitments mean that it falls to Neil Kessel to provide much of the unit's direction. Aubrey Lewis awards Kessel distinction in his diploma in psychological medicine, and Kessel works with Michael Shepherd in the General Practice Research Unit at the Institute of Psychiatry, where he delineates the concept of 'conspicuous psychiatric morbidity' – a psychological disorder known to a patient's GP.[4] He conducts studies on neuroses in general practice and alcoholism.[5] Kessel's work in Edinburgh is overwhelmingly based at Ward 3. He becomes professor of psychiatry at Manchester in 1965, where he remains for the rest of his career, assisting in the creation of a detoxification service for alcoholics, and becomes dean of the medical school and then postgraduate dean.[6]

Before the unit's transfer to Edinburgh, Kessel is not especially interested in attempted suicide; afterwards, in Manchester (from 1965), he focuses upon teaching and administration, also acting as government advisor on alcoholism (for the Department of Health and Social Security – the successor to the Ministry of Health from 1967). However, almost all of Kessel's work in Edinburgh concerns attempted suicide, and he proposes an important terminological shift: calling it self-poisoning. This interest coincides with Kessel's attachment to Ward 3 at the Royal Infirmary of Edinburgh. The physical/mental overlap enabled by the ward is most clearly shown in Kessel's previously quoted comments that the ward provides for patients who require 'overlapping general medical and psychiatric care'.[7] The addition of intense research scrutiny and national attention post-Suicide Act allows the object to flourish.

Institutional and national background

The institution of Ward 3 is, by the early 1960s, explicitly associated with the phenomenon of self-poisoning. It is seen to deliver a more or less complete sample for Edinburgh. Batchelor and Napier claim in the early 1950s that 'the large majority of all suicidal attempts occurring in the city of Edinburgh are admitted to this hospital', a claim which runs through almost all of their work.[8] Kessel's studies similarly argue that 'we observed more than 90% of all [attempted suicide] patients arriving at any hospital in Edinburgh'.[9] He is not arguing that the sample is representative of Edinburgh; more fundamentally, he claims that '[t]he case material is varied because it was complete'.[10] This coverage of Edinburgh is even said to obtain if poisoned patients are first admitted to another hospital, due to arrangements to transfer them to Ward 3. It is also stated more generally, that '[t]he emergency procedure for dealing with cases of attempted suicide in Edinburgh is widely known, simple to operate and rapid in its execution. It is invoked, on average, five or six times a week to admit a patient to ward 3 of the Royal Infirmary'.[11] Thus, practical arrangements – an established and well-publicised emergency procedure – allow the clinicians on a single hospital ward to speak of a city-wide phenomenon. Their claims to a complete rather than an arguably representative sample, mark Ward 3 as an exceptionally influential site of knowledge for attempted suicide. Kessel is cautious about projecting his conclusions beyond his Edinburgh sample (without questioning the completeness of that sample for Edinburgh). Other clinicians working at the ward see no reason for Kessel's caution, and claim that the Edinburgh figures are representative for Britain, 'as there is no reason to suspect that Edinburgh people behave differently'.[12]

The wider situation in Scotland is also noteworthy. It is brought up a number of times during reform campaigns for the suicide law in England and Wales that suicide is not a crime in Scotland.[13] When the law is changed, this does not apply to Scotland, and therefore neither does the hospital memorandum HM(61)94. It is notable that despite a standing rule for referral that is much older than the memorandum, there are very few studies of attempted suicide in Scotland until the impetus and publicity of the 1959 and 1961 acts.[14] One study presumably prompted by the legal shifts is the effort of A. Balfour Sclare and C.M. Hamilton in Glasgow – a study that is dwarfed by the institutional and research potentials at Ward 3. Based in the Department of Psychological Medicine at the Eastern District Hospital, Glasgow, most of the study's patients are referred from the Glasgow Royal Infirmary. Over half of the attempts are said to be motivated by either 'marital and romance difficulties' or 'family relationship problems'. In many cases the authors of the study characterise suicide attempts as 'a final act of exasperated abdication from what the patient regarded as an intolerable situation'. They do not see attempted suicide as a self-conscious appeal, but rather as a frustration reaction, a 'response to complex and overwhelming situations'.[15] Despite an established department of psychological medicine, this study does not appear to have an institution like Ward 3 to bolster its claims or a large number of full-time research psychiatrists and PSWs based at the hospital. Despite the Scotland-wide standing rule and the historic lack of legal constraint, this is just one more incidence of a growing problem across Britain.

Kessel's 'self-poisoning': similarities and modifications

Kessel's self-poisoning is different in three main ways from Stengel and Cook and Batchelor and Napier. The self-conscious nature of the appeal is the strongest and simplest notion of intent yet seen, and the archetypal behaviours and gender stereotypes are explicitly discussed. Further, Kessel's self-poisoning is rooted in an amorphous category of distress. This emotional state is thought common to all self-poisoning episodes, through which point of view it becomes a distinct, coherent clinical object. Thus in all three ways, Kessel's self-poisoning is more definitely, more precisely and more securely established: the intent is self-consciously to appeal; the stereotypes of young women and overdose are explicit; and interpersonal, present-centred stress and distress hold the object together at a deep conceptual level. However, much remains the same under this new term.

Much of the intense scrutiny focuses upon the familiar issues of lethality and intent. These function as part of a debate between therapeutic regimes. Kessel and two social-work colleagues make it very clear in 1963 that physical danger to life and psychiatric pathology are to be assessed separately:

> [N]o simple relationship exists between the degree of danger to life and the seriousness of any psychological disorder present. Many people who have been deeply unconscious we allow to go home after physical recovery because they require only a minimum of psychiatric supervision afterwards; on the other hand, a sixth of the patients who had not risked their life at all needed admission to a psychiatric hospital, and many more needed extensive out-patient care.[16]

There is a complex relationship between danger to life and a psychological disorder. Although they concede that 'on the whole…the more [physically] "serious" cases are more likely to call for active psychological intervention...it certainly is not right that mildness of method indicates lack of severity of psychological illness'.[17] A year later Kessel and various collaborators talk of the dangerous fallacy of 'using this yardstick of physical damage to judge whether the patient needs psychological treatment'.[18] The clinical object exists between therapeutic regimes, but (somatic) lethality is downplayed. The communicative overdose remains a tactical intervention between therapeutic regimes where the significance of the act is determined not by its physical consequences but its psychosocial context.

The first major difference is the explicit archetypal method. In 1965 Kessel entitles his Milroy (public health) Lectures at the Royal College of Physicians 'Self-poisoning'. These two articles are key in further publicising the terminological debate around attempted suicide. Rather than accept Stengel's and Cook's increasingly established modification of the term, Kessel finds attempted suicide, 'both clinically inappropriate and misleading',[19] advancing self-poisoning because it allegedly' describes the phenomenon without interpreting it along a single pathway'.[20] However, Kessel is opening up and closing down various possibilities. His terminological offering is intended to sidestep issues of intention ('interpreting' here indicates assessments of intent), but collapses all possible behaviours into one archetype.

These lectures describe a rather unorthodox practice in the promotion of self-poisoning stereotypes. Kessel sends an actor into six chemist shops in Edinburgh, instructed to simulate being in floods of tears, and to request two hundred aspirin. Kessel reports that she was served in every shop and only once was concern expressed: 'Two hundred? Are you all right? You ought to go and have a cup of tea'. This state of affairs is described in strong terms as irresponsible.[21] This expresses the twin facts that a 'sobbing girl' is typical of self-poisoners and that purchasing a large quantity of aspirin in this state is obviously a suicide risk – which also narrows the method of attempted suicide.

Kessel concedes that definitions of self-poisoning are difficult and fraught with complexity, but he uses phrases like 'mimicking suicide', 'simulation of death', and 'drama was enacted for their own circle'.[22] Such phrases expose a simplification of intent: this is not Batchelor's and Napier's childhood emotional trauma surfacing, nor Stengel's and Cook's unconscious, ambiguous ordeal. This is performance, deception and drama. The object is still unarguably social, but now very much self-consciously so. This is clearest in one of his last publications on the subject: 'The respectability of self-poisoning and the fashion for survival' (1966). He claims that 'it is common knowledge that you can take a lot of pills, lose consciousness and later return to it none the worse for the experience'.[23]

Alongside this stabilisation of intent, Kessel's self-poisoning is based upon the stereotypical method, the assertion that distress is the one common feature in all self-poisoners, and an effort to present this as a predominantly female behaviour pattern. All these have resonances and connections with wider trends during this period, in a different register to their resonance with the commitment to social welfare. Poisoning with drugs is linked to anxieties about prescribing and pharmaceuticals (although illicit substances do feature in a small way). Distress is a broad conceptual foundation for the turn to the social psychological medicine in this period. The gendered character of self-poisoning is linked to a feminised vision of domesticity through psychiatric social work. These three parts of Kessel's self-poisoning are explored in turn.

Poisoning, overdosing and drugs: local and national concerns

Kessel does not totally close off other behaviours possibly covered by attempted suicide (self-cutters or throat- or wrist-slashers, for example) but his terminology is exclusionary, even if those so identified are still treated at the ward.[24] Awareness of the phenomenon of self-poisoning with drugs increases during the 1940s and 1950s. According to one Edinburgh toxicologist: 'The first resuscitation centre dedicated to poisoned patients' opened in Copenhagen in 1949. In England, the North-East Metropolitan Regional Barbiturate Unit was set up in Romford in the 1950s.[25] Comments made in the late 1950s by the head of the Romford unit indicate that certain forms of poisoning have affinity (in the eyes of some clinicians) with suicidal gestures: 'barbiturate poisoning is notorious in that it is not a particularly lethal variety of poisoning[;] [it] is important because of its frequency and not because it is highly lethal'.[26] He does not comment further on the consequences of toxicological assessments of lethality for psychological assessments of intent. However, Stengel and Cook make a connection with poisoning in general, arguing that '[c]learly, the degree of danger to life is not a reliable measure of seriousness of intent, especially with poisoning, i.e. in the majority of suicidal acts'.[27] Thus, ambiguity of method is transposed onto ambiguity of intent, giving this method increased visibility. However, there is nothing inherently ambiguous about this method; such a claim falls into technological determinism. The explicit, conscious nature of the appeal in Kessel's self-poisoning overrides any ambiguity in the method in the above claim that 'it is common knowledge that you can take a lot of pills, lose consciousness and later return to it none the worse for the experience'.[28]

As we saw in Chapter 2, it is likely that the secure status of Ward 3 brings attempted suicides to prominence there, transferred for their own protection rather than because of any illegality (as attempted suicide is not an offence in Scotland). The Ward's poisoning associations shift from delirium tremens and alcoholism in the early twentieth-century through to attempted suicides, which is collapsed into poisoning and then broadened out to encompass accidental poisonings.

In 1962, a subcommittee of the Standing Medical Advisory Committee – under the chairmanship of Guy's Hospital Surgeon Hedley Atkins – issues a report titled: 'Emergency Treatment in Hospital of Cases of Acute Poisoning'.[29] According to a 1959 memorandum, this committee is set up on the basis that

> [a] certain amount of publicity is constantly being given to the dangers associated with poisons. Questions in the House of Commons recently expressed anxiety at the increase in accidental deaths due to the barbiturate group of drugs, and the Minister of Health said in reply that he would ask for attention to be paid to the need for special caution in their use.[30]

It is notable that (as late as 1959) accidental, rather than suicidal, poisoning is the reason for the committee's establishment. This report has specific significance for Ward 3, which in 1962 is designated a regional poisoning treatment centre (RPTC) in accordance with Atkins Committee recommendations. The Hill Report (1968), issued by a committee chaired by Denis Hill, reiterates the earlier recommendation that regional poisoning treatment centres should be established for the specialist treatment of acute poisoning.[31] Henry Matthew, Slater's successor as physician-in-charge of Ward 3, comments in 1969 that '[s]uch a centre has evolved at the Royal Infirmary of Edinburgh over the past 90 years, and during recent years it has functioned in the manner recommended in the Atkins and Hill Reports'.[32]

One of the papers circulated to the Atkins Committee in the early 1960s involves a more technical – but still important – concern: having ambulances carry the right mix of carbon dioxide and oxygen with which to treat patients poisoned with carbon monoxide. This shows how 'acute poisoning' is not necessarily associated with pills or an 'overdose', but during this period it becomes that way. The decline of carbon monoxide or coal gas poisoning – the method with which poet Sylvia Plath ends her life in 1962 – coincides with the increasing number of British houses switched from coal gas to natural gas from the mid-1960s – a trend studied in both Edinburgh and Birmingham.[33] These concerns show the narrowing that takes place when switching terminology from self-poisoning to overdosing – there is no normal dose of carbon monoxide, thus overdosing makes little sense as a description of this method.

The wider significance of the overdosing archetype is explicable partially in terms of anxieties around prescription medication. It is in this context that Stengel blames the increased availability and consumption of sedatives under the NHS for the preponderance of drug-based attempts.[34] Kessel agrees, claiming that '[s]leeping tablets, and they are mostly barbiturates, are the accepted mid-twentieth-century passport to oblivion, and doctors seem only too ready to issue the necessary visa'.[35] The importance of drugs as the archetypal method of communicative attempted suicide continues to rise throughout the 1960s. General Practitioner C.A.H. Watts expresses the opinion in 1966 that '[t]he death of Marilyn Monroe has no doubt helped to popularize the overdose of sleeping tablets. Suggestibility and fashion, together with the fact that from 1961 attempted suicide ceased to be a felony [*sic*], in part account for the incredible number of attempts which occur today'.[36]

Concerns around overprescribing are exemplified by Karen Dunnell's and Ann Cartwright's book, *Medicine Takers, Prescribers and Hoarders* (1972),[37] which is also part of the important and complicated issue of the supposedly meteoric rise of psychoactive medications in mental health care and the technologies of the randomised controlled trial (RCT).[38] In a non-psychiatric context, there is a huge crisis of confidence over drug safety around the Thalidomide disaster. During the late 1950s and early 1960s this drug is prescribed as an anti-emetic (among other things) to help to counter the morning sickness associated with the early stages of pregnancy; it is then causally associated with malformations of foetuses.[39] The committee set up to enquire into how this could have been allowed onto the market is chaired by Derrick Dunlop. Drugs register on still broader levels. Russell Brain's committee on drug addiction issues reports in 1961 and 1965 on morphine, heroin and cocaine addiction.[40] There are well-publicised debates around cannabis, and when the Wootton Report recommends the decriminalisation of cannabis in 1969, Home Secretary James Callaghan is sufficiently moved to speak out in the House of Commons against the 'advancing tide of so-called permissiveness' in the country.[41] In the midst this, Kessel's narrowing of a behavioural stereotype around attempted suicide passes almost unnoticed.

Distress and the social constellation

The second of Kessel's key modifications concerns distress, a concept shown in the introduction as having inescapably social overtones. He explicitly adapts Batchelor's and Napier's insights on the aetiology of this phenomenon, moving away from childhood emotional trauma towards present-focussed stressful situations. This shift can partially be explained through changing PSW practice. Furthermore, distress allows pathology to be projected onto individuals in the social setting rather than the patient admitted having self-poisoned – typically a pathologically jealous husband driving his otherwise normal wife to a suicide attempt. This development is also related to PSW practices, especially the influence of marriage guidance. Kessel is not the first to use the terms stress and distress to describe this phenomenon, but he is the first to unify it in this way.[42] He asks: 'Is there a unifying basis to self-poisoning acts? Is there some feature that informs them all?', then answering, 'Distress drives people to self-poisoning acts: distress and despair, unhappiness and desperation'.[43] Edinburgh PSW J. Wallace McCulloch and psychologist Alistair Philip declare in *Suicidal Behaviour* (1972) that '[w]e firmly endorse Kessel's statement that "*distress drives people to self-poisoning acts.*"'[44] It is explicitly emphasised at the core of the behaviour.

Distress functions in a similarly cohesive way to Kessel's earlier use of the term neurosis, where he claims that '[n]eurosis is an agreeably vague word used here to embrace all those emotional disturbances, anxiety states, hysterical reactions, phobias, obsessions and depressions which become transmuted into illnesses by the simple process of taking them to the doctor'.[45] In a similar vein, Richard Asher claims that 'an increase in illnesses caused by stress – the huge amount of psycho-somatic illnesses found today – [does not] mean anything more than a shifting of the blame for their troubles which both doctors and patients like to place squarely on some real or imaginary source'.[46] Asher does not see the increase in psychosomatic illness as part of a growing overlap between separate therapeutic regimes; rather, he attributes it to fashions in disease, just as Kessel talks of the 'fashion for survival' after self-poisoning.

There is a distinctively evolutionary angle to much work on stress. As we have seen, Walter Cannon and Hans Selye draw their insights from animal experiments, and stress is theorised as an adaptive response applicable more broadly to animals and rooted deep in the evolutionary past (otherwise animal results have no significance for humans).[47] What is interesting about the ideas of distress mobilised by clinicians concerned with attempted suicide is the lack of explicitly evolutionary explanations, the use of animal experiments and ethology. Clearly, the concepts of distress and stress gain traction because of these influential explanatory systems. It seems futile to deny that the stress described here might rely at one level on unspoken evolutionary assumptions. On one hand John Bowlby's theories – used to underpin attempted suicide studies in the 1950s – come to have significant ethological underpinnings, but Erwin Stengel's work moves from a position of ambivalence (in the 1950s) to outright scepticism (in the 1960s) about the deep evolutionary underpinnings of attempted suicide.[48]

On the whole, distress functions as a broad, under-theorised blanket explanation, uniting concerns about '[s]ubjectivity, meaning, idiosyncrasy, feelings, a social nexus'.[49] Whilst Rhodri Hayward has shown that George Brown's and Tirril Harris' work on stress and life events in 1970s Camberwell is underpinned by an appeal to an 'evolutionary context ... a familiar ethological drama of confrontation and withdrawal',[50] this emphasis is not overt analyses of attempted suicide in the late 1950s and early 1960s. In 1992, Raymond Jack surveys the models that have been used to explain self-poisoning. He argues that stress has been seen as key, and shows how closely this term comes to stand in for the social environment: '[S]tress is external to individuals and emanates from the social conditions which govern their everyday lives'.[51] Kessel's distress gains purchase through a rhetorical, all-encompassing self-evidence, which (as argued in the introduction) is necessary for psychiatric epidemiology and social psychiatry in order to make sense in the post-war period. This distress, bound up in conceptions of the social environment may be self-evident in certain contexts, but Kessel's is also rooted in PSW practices – part of the state's commitment to psychosocial management.

Social settings and social workers – PSWs at Edinburgh and beyond

During the early 1960s, PSWs occupy a prominent place in Kessel's studies. He works most closely with PSWs Elizabeth Lee, then J. Wallace McCulloch, continuing the collaborative focus of Batchelor and Napier and Stengel and Cook. According to MRC records, 'in Edinburgh the Medical Officer of Health was an enthusiastic exponent of home treatment for the mentally ill and had been training his Health Visitors to act as P.S.Ws. This was not true of the surrounding localities'.[52] The potential to carry out such investigations is not widespread. In fact, to have PSWs as part of a local authority service (as they would be if combining the role with Health Visiting) is exceptional.[53] The broad shift towards community care brings social work to renewed prominence. In a 1968 textbook of psychiatry for social workers it is claimed that '[p]sychiatry is showing a healthy tendency to emerge from hospital into the community and in doing so it leans much more heavily than before on the assistance of every type of social worker'.[54] As mental health care becomes increasingly organised around outpatient departments, the twin practices of home visiting and social history-taking have even more potential to fabricate a credible social space around any given case of mental disorder. There is thus a significant amount of socially focused expertise upon which Kessel can draw.

Despite the health visitor–PSW training scheme, Kessel complains in 1962 that a '[s]hortage of psychiatric social workers makes it difficult to obtain additional information; when their services are available it is more often to provide after-care than to augment the history'. However, a footnote acknowledges: 'This paper was submitted for publication in 1961. Since then there has been an increase in the allocation of psychiatric and social work time. This now permits a fuller investigation of each case'.[55] Difficulties elsewhere are hinted at by John Wing in 1963, when he describes some of the arrangements for a psychiatric research project in London: '[T]here will be three social workers involved. It is not usually possible to find highly qualified, trained people for this work'.[56] We have seen in Chapter 2 how PSWs at Edinburgh impact upon the knowledge produced about attempted suicide. They broaden the spaces of investigation, from the various hospital spaces (the accident and emergency department, Ward 3, etc.) through home visits and follow-up, enhancing the credibility of any projections into those spaces. These visions of domesticity help to stabilise this phenomenon.

Kessel is explicit (to a much greater extent than Batchelor) about the PSW role in the investigations into self-poisoning. In 1963 he argues that 'we need as much of the P.S.W.'s time as of the psychiatrist's, which 'reflects the importance we place upon social work both in elucidating the circumstances leading to the overdosage and in dealing with the complicated social nexuses and tangled personal relationships that beset so many of these patients'. He also notes that arrangements are made to interview a spouse or other relative (called a key informant), and this information is fed into a conference where 'social and clinical details are put together'.[57] These are the practices upon which an interpersonal, social constellation is built.

The role of the PSW in fabricating a social nexus around a patient is put into context by Noel Timms in 1964 when he notes that a 'considerable number of referrals by psychiatrists are still requests for a social history'.[58] Such histories are

> a most important element in understanding the patient and his illness... As we have seen, treatment in psychiatry is not solely concerned with the patient. It is concerned with the patient in his total environment which includes his family, his home, his work and all other areas of his existence that affect his mental well-being... it is necessary to learn a great deal about the patient's social constellation.[59]

This social constellation is not static. Changes are apparent during the 1960s as social workers are advised: 'Unless financial hardship is patently a factor in the patient's mental disturbance it is not usually necessary for the psychiatrist or the social workers to obtain minute details of family income and expenditure'. In addition, it is 'not enough to record the district or municipal ward in which the patient lives as an indicator of his social status' due to housing shortages, housing policy and increased social mobility. Instead, 'it is better to discover whether the patient is suited or unsuited to his home area and whether he and his family are happy to conform to the prevailing standards of the neighbourhood'. From implied previous concerns around poverty and fixed urban spaces (which are also traditional sociological concerns), the issue becomes one of adequate psychological adjustment within any given social environment: 'This account of the patient in his social milieu is a valuable background to the more detailed information on the patient's emotional environment which the psychiatrist will gather from the patient himself'.

Given McCulloch's interest in the subject, it is unsurprising that this co-authored textbook, *Psychiatry for Social Workers*, should accord a special place for social worker interviews around attempted suicide. It is noted that 'we [have already] described a schema for a standard social history, but in the case of attempted suicide there is a good deal of additional information which must be obtained before the significance of the attempt can be adequately assessed'. Munro and McCulloch set out a scheme for the recording of data for the specific occurrence of attempted suicide, which includes the patient's indications of their intent to others, the circumstances in which the attempt occurred, the measures taken to either ensure or avoid discovery and the reactions of relatives.[60] This is a revealing didactic practice for the consistent fabrication of a social environment around a presumed attempted suicide (rather than investigations of the patient's constitution or brain chemistry, for example). Kessel also sees the dramatic nature of self-poisoning as requiring PSW assistance. He claims that GPs confronted with the phenomenon 'will need the services of a psychiatric social worker, so that an informant's account can be obtained in all cases. Very often the patient himself will conceal important information ... so as to extract the last ounce of drama from a situation in which he holds the centre of the stage'.[61] The language of deceit solidifies the self-conscious character of intent and shows its reliance upon social work practices.

The present, marriage guidance and managing the boundary of pathology

Kessel's self-poisoning is rooted in the present. Joan Busfield argues that the relationship between stress and mental disorder 'focuses not on events in early childhood but on an individual's more immediate situation'.[62] Whilst stress is not inherently present-centred, Kessel's modification of Batchelor and Napier is of interest in this regard. In a paper published in 1965, Kessel and McCulloch use their concept of distress to modify Batchelor's analysis:

> Batchelor (1954) has suggested that those who act impulsively [when attempting suicide] are manifesting an acute frustration reaction and this aspect we recognize. But our impression is that they do it not so much because they are or feel thwarted as because they are distressed ... Distress, whether it stems from depression or from intolerable social circumstances, is always present at the time of the act.[63]

As we have seen, Batchelor's and Napier's work pivots upon an acute frustration reaction linked to childhood emotional trauma. This thread re-emerges in 1960s attempted-suicide studies from University College Hospital (see below). Kessel and McCulloch instead emphasise present distress over past emotional deprivation, and the present social environment against the childhood emotional environment. Kessel is also ambivalent about Batchelor's and Napier's reliance upon the concept of faulty adaptation: 'Whether the broken parental home is the root from which stems the disorganized life pattern ... must remain a matter for speculation'.[64] Kessel instead relies upon notions of impulsivity rather than frustration: 'Two-thirds of all acts were impulsive...This astonishing finding is of the utmost importance. Five minutes, sometimes only one minute, before the act took place the idea of taking poison was not in the person's mind'.[65] This is a clear shift.

This move towards the present shows psychiatric social work's expansion beyond child guidance into marriage guidance, a movement founded in the 1920s with significant connections with PSWs.[66] The Family Discussion Bureau is founded in 1948 by the Family Welfare Association and becomes attached to the Tavistock Institute of Human Relations in 1956. PSWs began to be trained in the 'psychology of family relations' from the late 1950.[67] These concerns also resonate within psychiatric research, for example Norman Kreitman's studies at the Graylingwell Hospital in Chichester in the early-mid 1960s into mental disorders and marriage. These studies draw upon the eugenic concerns of Lionel Penrose's study of 'Mental Illness in Husband and Wife' (1944) and Eliot Slater's and Moya Woodside's *Patterns of Marriage* (1951).[68]

This increasingly marital focus feeds into Kessel's present-centred distress. It is seen as 'the chief aetiological factor in many cases' and, in general, 'the attempt follows swiftly upon an acute domestic quarrel in a chronically disturbed matrimonial situation'. Batchelor's broken home is placed on an equal footing with the concept of a 'breaking home', which is present-focussed; the aetiology of the attempted suicide migrates from the past to the present.[69] Present marital disharmony is only a short step away from broader romantic, communicative interpersonal concerns. Kessel argues:

> Admission to the ward, having poisoned oneself, can be for instance a powerful weapon in bringing back errant boy friends. The girls who resort to it are, all the same, very much distressed; in their despair they do something stupid and senseless, and it works...Perhaps what we most resent is that, though there was probably a negligible risk to life, they are held by their circle of friends narrowly to have escaped death. They have had their drama; to us it only means work.[70]

The highly gendered nature of this communication is discussed below. For self-poisoning to be a powerful weapon it must be rooted in a present social context.

On a practical level, in 1964 Noel Timms sees slight but significant temporal changes in the social history: '[P]sychiatric social workers now think they are called on not so much for a detailed expression of family history but for an assessment of the present situation'.[71] More theoretically, PSW Eugene Heimler argues in 1967: 'In community care the present plays an extremely important part...the theory of psychiatric community care is this: the past influences the present, but the present also influences the past'.[72] Munro's and McCulloch's section on history-taking also shows the growing influence of the present. Under the PSW's heading, 'Home Circumstances,' should 'be described the circumstances which are typical of the patient's current life rather than those which were present in his earlier years'.[73] It is clear that longer-term factors can co-exist with this focus on the present, but the present-centred concerns of the mid-1960s throw the work of Batchelor and Napier into sharp relief.

This present-focused distress also forms part of a complicated relationship between abnormal action and psychiatric pathology. Kessel states: 'It has often been argued that to poison oneself is such an abnormal act that everyone who does so must be psychiatrically ill. We have not fallen into that tautological trap'.[74] The focus upon marital relationships also has a significant role in managing the ambiguously pathological nature of 'distress'. Regarding self-poisoners, Kessel continues: 'Of particular importance is the fact that 26% of the men and 20% of the women had no psychiatric illness.[75] The pathology does not disappear: marriage and the social constellation allow pathology to be projected onto somebody who has not even been poisoned. McCulloch and Philip put this most clearly in 1972:

> [T]he Edinburgh studies have shown that among married women pathological jealousy in the husband was found in almost a quarter of the cases. Indeed, the persistent suspicions of the 'jealous husband' were frequently found to be a precipitating factor for the attempt. In all but a tiny proportion of such cases, the husbands themselves reported that their jealousy had been completely unfounded.[76]

This idea of illness emerges right at the point where marriage guidance and psychiatry intersect. The figure of the jealous husband is given an entire chapter in J.H. Wallis's text, *Marriage Guidance: A New Introduction* (1968). Wallis ends his description with: 'The important question [is] whether this client may need psychiatric treatment', and he refers to that same problem: 'There cannot be a categorical answer to this question since the dividing line between sickness and health is not precise. One has to consider the whole situation'.[77] The social setting, psychiatric treatment and the boundary between mental health and illness link psychiatrists, PSWs and marriage guidance counsellors around this object of self-poisoning.[78] The marital relationship is subject to intense psychiatric scrutiny through interviews, follow-up and case conferences.

Distress, domesticity and gendered self-poisoning

These practices are saturated with stereotypes of femininity. Nevertheless, this is a highly uneven gendering process, left unexplained or unmentioned; as Raymond Jack rightly points out the issue has 'been virtually ignored in the literature'.[79] There is certainly nowhere near as much crude gender stereotyping as that which pervades the late 1960s North American–based stereotypes of delicate self-cutting, which begin to seep into British practice by the middle of the 1970s (see Chapter 5). All three of Kessel's modifications (self-consciousness, poisoning and stress) have potentially gendered freight.

The additional self-consciousness feeds into stereotypes of feminine manipulation, exemplified by Kessel's above-quoted comment about bringing back errant boyfriends. Self-poisoning is also seen as a passive (read: feminine) method, which interacts with a gendered imbalance in the prescription of barbiturates. Ali Haggett states: 'Since the 1970s, feminist historians have suggested that the lack of opportunities afforded to women and the banality inherent in the domestic role caused symptoms of anxiety and depression in post-war housewives. Correspondingly, they have argued that the primary motive for prescribing psychotropic drugs was to ensure that women "adapted" to their domestic role'.[80] Finally, distress has resonances with supposed feminine emotionality, but is also explicitly articulated as part of this feminised domestic role.

The projections enabled by psychiatric social work practice, principally around 'distress', interact further with marriage concerns in a domestic-centred way. Indeed, Kessel makes 'the emotional' a cornerstone upon which he can build a 'domestic space' in this fascinating (and explicitly normative) gendered passage:

> There is no simple explanation of the high rate of self-poisoning among young women in their early twenties...These women, although fully engaged in their normal social setting, mothering and running a home, are emotionally isolated...they have not yet had time to adjust to the confines of domesticity...Unhappiness mounts, and then suddenly explodes, at a moment of special crisis.[81]

This recalls Slater's and Woodside's home interviews of soldiers' wives in the late 1940s, where Woodside reports witnessing 'struggles and ambitions eventually adapting themselves to the limitations of a restrictive environment'.[82] Indeed, marriage, domesticity and psychopathology are historically well-connected.[83] This general emotional isolation and supposedly normal social setting are opened up for Kessel through PSW spouse interviews.

We noted one phenomenon over and over again. An insensitive spouse, generally the husband, although he cared for his wife had failed to notice either her need for emotional support and encouragement or the growing sense of isolation within the home that stemmed from their lack.[84]

Domestic stress is still gendered, not through Bowlbian maternal deprivation but through a feminine lack of resilience, or a masculine lack of support. These gendered gaps affect Kessel's way of framing and answering questions: 'Confirmation was thus provided of the clinical impression derived from dealing with the patients, especially the women in the ward, that marital conflict is the chief aetiological factor in many cases'.[85] The practice of holding a clinical conference with PSWs at Ward 3 has been made a rule by February 1963.[86] This co-operation brings in credible information, accessed by interview with somebody who is not a patient, opening up a space where Kessel's casual clinical impression can gain empirical validation or confirmation. Thus, he is able to speak about domestic space through what is observed in a hospital ward. Once this clinical impression is confirmed, it can predominate, even to the point of overriding PSW input that helps to enable it: 'The psychiatric social worker, who had seen both partners, graded only half the marriages as poor or bad ... Perhaps, however, one has to be inside a marriage really to assess its satisfactions and its failures'.[87] Visions of the home are created in these analyses, as part of the wider project that inscribes mental health and mental disorder onto the social, interpersonal fabric of everyday life.

The unequally gendered archetype is tackled explicitly by Kessel, who disagrees that self-poisoning is 'the female counterpart of delinquency in young men ... [which] would suggest that women turn their aggression against themselves, while men act against society'.[88] He argues, instead, that self-poisoning is better understood through emotional isolation and failure to adapt to domesticity. Through his rehearsal and rebuttal of a delinquency hypothesis, Kessel explicitly demonstrates a move away from conventional, significantly masculine, sociological concerns (such as crime, delinquency and deviance), to a position made possible by the PSW-founded analysis of domesticity. This is a crucial component of his rendering of female-dominated self-poisoning. But it is not enough merely to state (and lament) the traditional association or, more precisely, mutual constitution of domesticity with femininity. Sexism is active practice, not merely a re-articulation of established associations.

Psychiatry, the social setting and women are closely connected during the 1960s. The influential *Psychiatric Illness in General Practice* (1966) goes so far as to say, '[I]t would be a justifiable exaggeration to say that in the eyes of the general practitioners, psychiatry in general practice consists largely of the social problems of women'.[89] A gender imbalance in communicative overdoses does not seem exceptional in the wider context of reading mental illness into interpersonal relationships. The idea that those gendered female are physically, emotionally, psychologically or evolutionarily more suited to domestic, home or family spaces, is a durable plank in circular sexist arguments that feminise domesticity *a priori*. This gendered imbalance is rooted in understandings of home, as child and maternal bonds receive an increasing level of criticism after the mid-1960s. As Rose argues:

> In the 1940s and 1950s those who rallied round the cause of mother-hood and deprived children considered themselves progressive and humanitarian, in touch with the latest scientific evidence on the nature of the family... But in the mid-1960s this amalgam of theo-retical systems professional practices, legislative measures, social provisions, and public images – this 'maternal complex' – came under attack. Historians and sociologists challenged the universality of the mother-child bond, and hence its claim to be 'natural'... Feminists criticized it as little more than a means of enforcing and legitimating women's socially inferior position and their exile from public life.[90]

During Kessel's time at Edinburgh, such critiques are far from the main-stream and, even afterwards, they struggle to make much headway in psychiatry. However, this 'maternal complex' is another part of the social commitment that is rolled back in the 1980s. Additionally, the move from past to present – from broken homes to pathological marriages – enables a specifically feminine aspect to self-poisoning to emerge. Broken homes affect both genders more or less equally, but this is not the case for present domestic problems. This reassertion of gender difference is connected to an increased reliance upon social work, which has a gendered dynamic of its own.

John Stewart notes that during the interwar period, 'social work was...a predominantly female occupation',[91] an assessment echoed by Noel Timms in the post-war period.[92] Of course, the presence of those gendered women in any given profession does not necessarily mean that the work produced will be gendered in any particular way. The problem arises from the gendered assumptions that are articulated through the imagery and associations of a supposedly female profession. The child-guidance roots of PSWs carry significant gendered freight, and Timms is aware of the gendered belittling of PSWs by psychiatrists. He recalls an article in the *BMJ* in 1950 on 'The Role of the Psychiatric Social Worker':

> Dr J.B.S. Lewis appeared to give full recognition to the psychiatric social worker. 'She should', a report of the meeting states, 'of course, work in close conjunction with a psychiatrist; but it must be remembered that she had a skill of her own, and he could learn from her as she from him. Her duties were multifarious. She had to explain to the patient, his relatives, employers, etc. what the hospital or clinic was doing; to take a social history; to follow-up and help discharged patients; to co-operate with other social services; to help in administration and therapeutic work and in research; and, in fact, to carry out many *other chores*'.[93]

This earnest and patronising picture is assessed with Timms' sardonic comment: 'The fairly high status accorded to the psychiatric social worker is somewhat diminished by the ambivalent comment in (my) italics'.[94] Scrutiny of domesticity is elided into domestic work (chores). The sexism upon which pathological domesticity is founded is the same sexism that saturates the profession of psychiatric social work. In all of Kessel's moves, from self-poisoning to self-consciousness to domestic distress, the gendered character emerges, hand in hand with a patronised profession of PSWs sent into the home space to bring it back for the psychiatrist's reimagining.

The various assumptions and methods of sense-making in this transformative expertise (including sexism, marriage guidance, and focus on the present) are inextricable from 'attempted suicide'. This phenomenon of 'attempted suicide' is a prominent expression of, and driver for, the broad and eclectic turn to 'the social' in mental health, which falls away as internal emotional regulation and neo-liberalism rise in the

1980s, laying the ground for biologised understandings of self-harm as self-cutting. The practical arrangements carried out in hospitals in the mid-to-late 1960s show how the psychiatric epidemiology MRC Unit is just a particularly bright spot in an increasingly varied field. Kessel is influential, but the phenomenon is on a much larger scale. However, this also brings significant problems outside of such established and insulated therapeutic mixtures as Ward 3.

Observation ward to DGH unit: practical integration and new crossover

After the Mental Health Act, the equation of mental with physical illness enables mental health care on the same deregulated basis as physical care. In practical terms, the integration of psychiatric with general medicine is attempted by casualty referrals, as we have seen, and the provision of psychiatric treatment units in DGHs. These units owe much to observation wards – in many cases, the wards become treatment units. Martin Gorsky argues that these units emerged in the 1950s and John Pickstone sees a tendency towards this kind of provision in the 1960s.[95] Walter Maclay goes so far as to claim that this 'new' trend for psychiatric units in general hospitals 'is really the reestablishment of an old pattern... In Scotland, general hospitals treated patients until the latter half of the 19th century'.[96] C.P. Seager claims in 1968: 'There have always been a large number of patients suffering from psychiatric illness treated in general hospitals. For a long time a large proportion of these were there by accident'.[97] Now their treatment there is self-consciously attempted.

These units are a key plank in the government policy of scaling back mental hospital provision. The *Hospital Plan* states: 'It is now generally accepted that short-stay patients should be treated in units nearer to their homes than is generally possible with large, isolated mental hospitals, and that it will usually be desirable to have these units attached to general hospitals'.[98] One clinician observes in 1963 that '[w]hatever views may be held regarding the role of general hospital psychiatric units, they are increasing in number and influence, and their further development is accepted Ministry of Health policy'.[99] A team of clinicians at King's College Hospital (KCH) note in 1966: 'The Hospital Plan for England and Wales has made provision for a considerable increase in the number of short stay psychiatric units which will usually be attached to general hospitals'.[100]

Psychiatric literature during the late 1950s and early 1960s is full of comment upon these local and specific developments.[101] Maclay argues in 1963 that 'psychiatric outpatient work should be carried on in the general hospital even if there is a nearby mental hospital ... this is vital if psychiatry is to be integrated with general medicine'.[102] The desirability of these units goes beyond spatial advantages, and is far more about the administrative isolation to which mental medicine is still subject.

Observation wards frequently become DGH units. Freeman notes that '[m]any of these [observation ward] facilities were later to become general hospital psychiatric units, particularly in Lancashire'.[103] This also happens in London in the former observations wards at St Pancras and St Clements.[104] D.K. Henderson argues in 1964 that observation wards 'paved the way for the more highly specialised psychiatric clinics'.[105] From Brighton, R.P. Snaith and S. Jacobson concur in 1965: 'As there are to be short-term psychiatric treatment units in general hospitals, we believe that much of the experience gained in observation units is going to be of inestimable value'.[106] The move from observation wards to DGH psychiatric units focuses attention upon the unhelpful stigma of segregated mental treatment. However, this undercuts the standing of the remaining observation wards, which go from embodying the integrationist and destigmatising spirit of the Mental Treatment Act (1930) to being overtaken by the 1959 Mental Health Act. Due to observation ward's secure and segregated nature and its enduring association with the Poor Law, it is undercut as a preferred method of crossover between psychiatric and general medicine.[107]

Manchester clinicians comment on the stigma of general hospital mental wards as early as 1949.[108] After the 1959 Act, such wards are even more out of step with the proliferation and integration of psychiatry through their differentiation between psychiatric and general patients. Stengel comments that the transfer of all attempted suicides to observation wards is largely 'impracticable, questionable on psychiatric grounds, and usually unnecessary. The practice is certainly out of keeping with the Mental Health Act 1959, which discourages discrimination against patients in the general hospital on the grounds that they present psychiatric problems'.[109] Observation wards become reconstituted as treatment units, or are replaced simply by having psychiatric beds on general wards. Psychiatric scrutiny becomes more diverse and subtle in its integration with general hospital practice, but also less protected by institutionalised arrangements. The eclipse of the long-established observation ward by new DGH psychiatric treatment units is a substantial change, and it provokes new conflict between therapeutic regimes.

The range of clinical phenomena coming to psychiatrists' attention in a general hospital is different from those in a psychiatric hospital.

There is awareness that this will change the kinds of clinical objects that emerge, as in a 1969 discussion of psychiatrist–physician liaison: 'psychiatrists who had not previously worked in collaboration with physicians in a general hospital clarified for themselves that they were called on to examine and treat cases differing from the range presenting in psychiatric hospital practice' which include 'personality disorders of moderate severity, resulting from disturbances in the patient's parental family relationship'.[110] The significance of the social setting again emerges under these new arrangements.

Separated therapeutics, beds and referral

These units are not without conflict. Despite – or perhaps because of – closer spatial integration, the therapeutic conflicts that undercut cooperation become sharper. Psychiatry and general medicine remain separate in this period, involving dissimilar, sometimes incommensurable, therapeutic approaches. The lack of administrative differences between them exacerbates friction between therapeutic approaches. This is not a problem exclusive to the post-1959 period. Back in 1953, R.W. Crocket at the Department of Psychiatry in Leeds wonders whether 'there is an inevitable conflict here, and that to combine the qualities required for first-class psychiatric care with those demanded by modern physical methods of investigation is an almost impossible achievement'.[111] There is abundant acknowledgement of therapeutic difference throughout the literature in the early 1960s, coupled with a sense that this difference is being lost or ignored in the headlong rush to proclaim psychic and physical ailments completely equal. A *Lancet* lead article puts it bluntly in 1962: 'The process of treatment is not the same in predominantly mental disorders as it is in predominantly physical ones; and this is something that must be made perfectly plain'.[112] Walter Maclay cautions in a similar vein that 'we must not lose sight of the basic truth that the nature of mental illness is different from the ordinary run of medical and surgical illness.[113]

Despite this enduring difference, psychiatric access to general wards increases – for psychiatric consultants for example.[114] However, whilst psychiatric units might be close by and even wards might be mixed, the basic unit of resources in a hospital, the bed, is still something largely – though not exclusively – subject to one set of therapeutic and diagnostic practices. Hospitals are predominantly made up of mutually exclusive 'beds' for various specialisms: geriatric, paediatric, psychiatric or surgical. Thus to produce a psychosocial context around a physical injury arriving at casualty – possibly also going for surgery or specialised resuscitation – requires referral to negotiate between these mutually exclusive spaces. Separation endures, as the walls of the asylum give way to the resource politics of mutually exclusive beds, an exclusivity founded upon ideas of therapeutic incompatibility. Nothing in the following section argues that somatic assessment or therapy is unnecessary. The argument is simply that because of the ways hospitals are set up with therapeutic approaches so separate, the priority of general, acute somatic medicine creates obstacles that need to be negotiated for a psychosocial attempted suicide to emerge.

Studies from Brighton, Leicester, Sheffield and Bristol, as well as several reports from an accident service at King's College Hospital (KCH) show how psychiatric scrutiny becomes reconfigured in general hospitals and how somatic medicine remains the primary concern in these environments. The practice of referral is the most important aspect of maintaining significant psychiatric scrutiny upon general hospital patients. However, varied practices are employed in DGHs to negotiate the therapeutic separation, practices that impact upon the psychosocial disturbance constructed around a presenting 'physical injury'. The Sheffield and KCH studies will be considered in detail below.[115]

Parkin and Stengel in Sheffield (1965)

One of Erwin Stengel's first major research projects at Sheffield (having been awarded the chair in psychiatry in 1957) is a collaboration with Dorothy Parkin published in 1965. The aim is to combine 'attempted suicide' numbers from three administrative levels (general hospitals, mental hospitals and general practice) into one composite incidence statistic. This study is based upon records rather than clinical encounters, but referral practices between therapeutic regimes are still vital.

The general hospital group comes from three Sheffield General Hospitals. However, 'attempted suicide' does not appear on casualty records. Although Stengel and Parkin claim that 'as a rule it was easy to pick out the suicidal attempts from the records', it is admitted that '[a]ttempted suicide is not a diagnosis and therefore does not appear in the diagnostic index of hospital records'. Instead, they use the following somatic categories recorded in casualty which 'served as indications for closer study of the casualty to which it refers: (a) no diagnosis, (b) collapse, (c) coma, (d) head injury, (e) laceration of throat and wrist, (f) stab wound, (g) poisonings of all kinds'. These somatic categories are transformed by closer study from Stengel and Parkin. The somatic therapeutics of casualty are thus further negotiated by referral to an on-call psychiatric consultant.[116]

Patients who are admitted end up at the psychiatric departments of these hospitals, 'transferred ... after the state of medical or surgical emergency had subsided'. Thus there are a number of ways through which these cases come to be labelled as attempted suicide. There is close study of certain somatic categories on casualty records; there is an on-call psychiatrist for those not admitted as inpatients; and there is referral to the psychiatric inpatient department once any medical or surgical emergency has been dealt with. In all these ways, somatic is transformed into psychological concern, negotiating the predominance and separateness of somatic therapeutics. They also note that 'in the psychiatric department of the Royal Infirmary a simple questionary is filled in for every new inpatient and outpatient. One group of questions refers to suicidal attempts'.[117] Thus, with a tick in the right box, a running record of attempted suicide is kept; put another way, a bureaucratic space is cleared, into which, at the stroke of a pen, cases arriving at certain departments of certain hospitals become conceptualised as 'suicidal attempts', rendered epidemiological and countable. Bearing in mind both Kessel and Stengel's points that '[a]ttempted suicide is neither a diagnosis nor a description of behaviour'[118] and will not show up in diagnostic records, such recording processes must be created, so that it might be inscribed, tabulated and transformed into a credible object of research.

The negotiations in the general practice group are different. Parkin and Stengel are open about these difficulties, noting that '[t]he size of the third group – that is, of those seen by general practitioners first – can be established only by a special survey'.[119] This GP input is carefully managed. The second question, 'How many patients did you *suspect* of having made a suicidal attempt?' requires clarification because 'doctors not versed in psychiatry and unfamiliar with the suicide problem tend to classify among suicidal attempts only those patients who admit suicidal intention'. The GP is compared unfavourably with the 'experienced psychiatrist [who], when seeing such patients in hospital does not find it difficult to elicit suicidal intention from them, or at least the feeling that "they did not care whether they lived or died." Many, perhaps most, suicidal attempts are carried out in such a mood'.[120]

This is an intervention designed to make the arena of general practice and that of the general hospital equivalent. It does this by using suspicion as a practical approximation for psychiatric expertise. This is something of an heroic effort at maintaining the attempted suicide with a stand-in for psychiatric scrutiny. Parkin and Stengel are perhaps aware of the stretch that they are asking their readers to make, as they add that a 'discussion with a group of general practitioners about the inquiry suggested that the inclusion of this question served the intended purpose'.[121] So whilst psychiatric expertise is not strictly essential to the production and maintenance of attempted suicide, significant intellectual labour to bring about an approximation is necessary.[122]

So whilst it may seem that general practice, or primary care, has been neglected in the wider story about the epidemic overdosing, it is simply that the organisation of health care in Britain makes it difficult and unlikely for attempted suicide to come under extended scrutiny in this area. C.A.H. Watts admits as much in 1966 when he comments that whilst '[t]he family doctor with psychiatric training may be able to deal with some cases [of attempted suicide]' what happens in practice is that 'most of the cases reported to us in general practice are seen at the time of the incident and need to be admitted to hospital for emergency measures, so they pass out of our care'.[123]

King's College Hospital Accident Service

There are six published reports from King's College Hospital (KCH) between 1966–9 either based around or with significant mention of attempted suicide. KCH has extensive geographical and practical links to the Maudsley.[124] P.K. Bridges and K.M. Koller (psychiatric registrars) and T.K. Wheeler (senior house officer) publish an account titled 'Psychiatric Referrals in a General Hospital'. They comment that a 'large part of the work in this department is concerned with patients who have attempted suicide', mentioning a 'regional accident service that has been developing in recent years[, and] which may partly account for the rising intake' of such cases. It is also argued that '[f]ollowing recent changes in social attitudes, suicide attempts appear to be increasing and it is likely that more of these patients now come to hospital'. There is also a rather opaque reference to 'increasing medical awareness of the potential significance of the suicidal attempt', which means that 'virtually all cases are referred to a psychiatrist'.[125] Bridges and Koller use the accident service in 'Attempted Suicide: A Comparative Study' in conjunction with a control group. The accident service is not specifically intended to bring attempted suicide into view but, due to this arrangement, there is a new potential field for clinical and research objects constituted through referral after somatic assessment: 'Virtually all cases of attempted suicide admitted to the hospital are referred for a psychiatric opinion'.[126] Bridges's 1967 remarks (from University College Hospital in North London) show the difficulty of establishing referral in accident departments, arguing that 'psychiatry has insufficiently been accepted into the general hospital and, therefore, Casualty Departments, where the need can be most acute, usually have considerable difficulty in obtaining psychiatric advice when it is required'.[127]

Interested in this phenomenon in his early career, H. Steven Greer signs a 1969 letter to the *British Journal of Psychiatry* that first proposes the term parasuicide (alongside Norman Kreitman and psychologist Alistair Philip from the Edinburgh MRC Unit, and Christopher Bagley from the MRC's Social Psychiatry Research Unit at the IoP.).[128] In 1966, when lecturer in psychological medicine at KCH Medical School, he reports on attempted suicide, with Koller featuring again, and also J.C. Gunn (a psychiatric registrar based at the Maudsley). They again mention the accident service, coupled with referral as key: 'Any patient who has made a suicidal attempt, however slight the medical danger, is admitted and referred for psychiatric opinion'.[129] This explicit mention of medical danger suggests the lowering of a threshold normally required for admission to the casualty department, and thus this arrangement helps to constitute a new field, at a casualty department, in which gestural suicidal attempts are more likely to become objects of scrutiny. It also functions to downplay the significance of somatic assessments, so that all patients come under psychiatric scrutiny, not just those coded (by physicians or surgeons) as seriously injured. The fact that 'medical danger' is self-consciously disregarded as a criterion for admission shows how 'gestural' injuries potentially might only become visible to psychiatrists at general hospitals because they are sought.

John Bowlby's ideas of childhood psychopathology re-emerge as Greer and colleagues explicitly question these attempted suicides about childhood parental loss ('broken parental homes') and any 'recent disruption of close interpersonal relations'. This is done through standardised practices, designed to result in a coherent object of 'attempted suicide': '[a] protocol was designed for recording relevant data about each patient. Information was obtained from structured interviews with patients, and in some cases relatives were also seen'. Through this they are able to claim that 'parental loss contributes to attempted suicide' as it 'predispose[s] to disruption of interpersonal relationships, and…childhood experience may make individuals abnormally vulnerable to the loss of a loved person later in life, thus precipitating suicidal reactions'.[130] This predisposition (based on faulty childhood development) is a key conceptual plank enabling past or present social environments to cause attempted suicide. In another study undertaken by Greer and Gunn only, patients from 'intact homes' and those who had suffered 'parental loss' are compared.[131] Thus, people are placed within a psychological nexus of childhood experience and interpersonal relationships. The conceptual apparatus of Bowlby, models of psychological development and patho-logical reactions to stress are by no means less important than admin-istrative and practical arrangements. Indeed, conceptual and practical labours do not occur independently of each other.

Unsurprisingly, given his previous work with Stengel, Kreeger's work on 'attempted suicide' at KCH is specifically focused upon these kinds of interpersonal disturbances. His approach is based on the principle that '[i]n every patient an attempt should be made to identify the nature of the appeal, whether this is for amelioration of environmental stress or for protection against overwhelming internal conflict'. He further claims that '[a]n attempt to understand the suicidal reaction in the context of the patient's life situation should always be made'. He adds that a joint interview is helpful in this process, bringing the relatives and social constellation to prominence: 'A joint interview with the patient and relative may reveal aspects of the relationship not otherwise apparent, as depressed patients are often unable to express criticism or even perceive fault because of their guilt and self-reproach'.[132]

Finally from KCH, J.P. Watson (based at St Francis Hospital) also uses the Accident Service to construct a series in which 47–53% of patients present with a 'suicidal problem'. A case 'was deemed "psychiatric" if the patient came to hospital with a problem relevant to psychiatry and did not require medical, surgical, gynaecological or dental treatment'. Thus psychiatry is defined, in practice, largely by the absence of other specialist attention. However, in psychiatry, one exception is made. The above definition comes with the significant qualification: 'unless he had deliberately poisoned or injured himself'. So psychiatric problems are normally accessed if there is no other claim on a patient in the general hospital environment. The exception is the self-poisoned or self-injured patient, where it is accepted that these patients might be treated 'medi-cally' or 'surgically' first. This shows once again how attempted suicide emerges through practices that negotiate the separated therapeutics of the district general hospital, in casualty departments.

So despite the best efforts of the *Hospital Plan*, therapeutic approaches remain significantly unmixed in this period. A number of different tactics, arrangements and procedures are necessary for attempted suicide to emerge. Some, such as Parkin's and Stengel's study, are designed to elicit an attempted suicide object, whilst still relying upon much wider systems of referral. Others, such as the KCH Accident Service, bring an attempted suicide to attention that is no less the product of human administrative intervention. Referral stands at the centre of these processes, right at the core of attempted suicide, the key enabler for the transformation of a presented physical injury into a psychosocial disturbance. However, there are noted problems around the practice of referral, and one of them is a conflict over resources between general hospital psychiatrists and other established specialisms such as surgery. These conflicts are useful when analysing how psychological, behavioural, clinical objects become established and self-reinforcing.

Social spaces embedded and established through the politics of therapeutic conflict

The final part of this chapter looks at how therapeutic conflict (rather than simply separated therapeutics) provides extra impetus for the establishment and entrenchment of a social constellation – specifically psychopathological domesticity – around a hospital presentation of attempted suicide. The increasing presumption of domestic psychopathology illustrates how behavioural objects become established. The social constellation, the domesticity fabricated by PSWs, appears stable and reliable enough to be presumed around physical injury. Psychiatrists report feeling pressure for a quick discharge of attempted suicides from general medical beds after somatic injuries have been dealt with. In response, the social constellation is increasingly invoked as a reason to keep a patient admitted. Thus the social setting shifts from being produced (laboriously) around an attempted suicide, to being deployed tactically in order to promote and sustain such scrutiny. The object becomes self-confirming, as the more obvious the act's communicative nature become, the more effort is expounded to discover a communicative motive. Finally, the object becomes a socially embedded, increasingly available option for the expression of distress.

The conflicts over admission, management and discharge are most explicit in Irving Kreeger's paper on the assessment of suicidal risk. He reports that a 'hazard arises when patients are seen in general hospitals after making suicidal attempts. There is usually considerable pressure for quick discharge ... from physicians, who resent their beds being blocked'. He places dramatic emphasis on the '[t]he irrevocable consequences of mistaken judgment [that] colour every aspect of our handling of the suicidal patient', with special emphasis on 'whether to treat a new patient as an inpatient or an outpatient'. This is a clear intervention in a conflict over scarce resources (beds). One of Kreeger's key arguments concerns the social environment that he, Stengel and Cook work so hard to establish during the 1950s, now deployed as a potential danger to the patient unthinkingly discharged. He emphasises that the 'patient can be at hazard for a number of reasons', including relatives in denial about the attempt, those too weak to support the patient, and those implicated in the cause of the attempt in the first place.[133] Whilst these assessments may push towards inpatient admission (to a psychiatric bed), it is also part of an explicit and concerted strategy against general physicians' pressure to discharge. Clinicians in Leicester bear this out: 'Because of the demand for beds' patient stays are 'generally too short for full psychiatric assessment'.[134]

Bridges, Koller and Wheeler also note serious pressure on resources, but suggest a more amicable resolution. Perhaps because psychiatry is well-established at KCH they are pleased to report that '[c]onsiderable co-operation was obtained from other departments so that many of the inpatient referrals received complete psychiatric treatment in a medical or surgical bed'. However, they complain that they have 'very few psychiatric beds', and that it is 'somewhat unsatisfactory' to use general beds for these patients. They are diplomatic, relating that '[t]here is always understandable pressure from physicians and surgeons for these patients to be transferred or discharged as soon as possible to allow further use of the bed', but resource conflict looms large. In this wider context they argue for a minimum of three days' observation for most patients so that 'the mood can be more accurately assessed, a social history may be obtained and the visitors may have facilitated the resolution of crises'.[135] Crucially, there are not only practical factors advanced in favour of continued occupancy of the bed (mood assessment and social history-taking), but visitors (cast as the social circle) are deployed as a reason for keeping a general hospital bed occupied by an attempted suicide patient. No amount of extra resources or efficiency in psychological assessment can speed up this visiting process that helps resolve crises. This is the precise opposite of Kreeger's thesis, but deployed in the same cause. Here the social generation and therapeutic repercussions of an attempted suicide become subtle but effective insulation against discharge pressure from physicians and surgeons.

Kessel's potentially psychopathogenic social constellation works differently again, maintaining a base for psychiatric credibility within the general hospital, but it is no less embedded through the tactical battle between therapeutic approaches. He and McCulloch (imagining the plight of other hospitals) clearly show how the pathological domestic situation calls for inpatient admission (which produces a need for further psychiatric beds):

[P]eople who poison or injure themselves are brought to hospitals and the physician or surgeon calls for psychiatric help. After physical recovery, if admission is needed to remove patients from an explosive domestic situation this will have to be to a psychiatric bed. Asylum is not a word psychiatrists use much nowadays, nor are they keen to bestow it. Yet many of these patients need a temporary refuge.[136]

Psychiatric credibility and the claimed necessity for further scrutiny are based on a vision of domesticity created by that very scrutiny. Kessel's and McCulloch's 'explosive domestic' situation, having been enabled by specific PSW practices, is now abstracted to general relevance in a claim on scarce resources. Instead of arguing for extended occupation of a general bed, Kessel and McCulloch call for more psychiatric inpatient space in a general hospital. Thus practical, tactical, resource concerns have a crucial role to play in the systematic emphasis placed upon the social constellation around an attempted suicide. These constellations are substantially sustained by politicking across the well-maintained split between general medical and psychiatric therapeutics. The production of a potentially psychopathogenic domestic space plays a key role in claim-staking in a general hospital environment.

'Splitting and Inversion' and established patterns of behaviour

It is precisely the success of the establishment of this attempted suicide that means the social constellation can be used in such conflicts. The consistent transformation of physical injuries into symptoms of a social constellation means that the latter (social constellation) can be used to explain the former. This is a gradual process occurring throughout the post-war rise of this epidemic phenomenon. In rather technical, esoteric terms, the success of these practices allows the social constellation to be 'split and inverted', becoming productive of attempted suicide. The mechanics of this process are well explained by Roger Krohn, who draws upon Bruno Latour and Steve Woolgar to claim that 'the constructing sentences are split from their imaginary objects, and then the now real objects are assumed to have caused the sentences'.[137] Krohn is talking about images and diagrams, but this is a useful concept to explain how referral, PSW interviews and psychiatric scrutiny being brought to bear on patients first encountered in a hospital can be used to create a pathogenic social space.

A patient arrives at A&E with a certain kind of injury (e.g., poisoning), possibly unconscious or semi-conscious. After somatic treatment (possibly stomach washing), practices of referral are required in order to question and assess the patient from a psychological point of view. Somatic treatment does not require an extensive reconstruction of the precipitating or family circumstances. However, this is the principal aim of psychological scrutiny – to produce a social situation once a physical injury has been referred for assessment. This situation then gets 'split' from the practices that produce it and inverted so that it is positioned as prior to the episode, and can now cause it. This is possible because social stresses (present) or predisposing factors (past) act as a conceptual bridge between circumstances and a behavioural pattern. Hence, statements that marital disharmony or broken homes cause self-poisoning are possible when viewed from a hospital ward. Once this process begins to recur predictably, the positioning is not so simple: the practices and the projections become mutually constitutive.

It is at this point of mutual constitution – when meanings and pathogenic social spaces are established, to then be deployed to reinforce the scrutiny that produces them – that the object can be considered established. This self-reinforcing process can spread and, to paraphrase Hacking, new possibilities for action become a culturally sanctioned way of expressing distress.[138] However, as has been argued here, this concept of distress is linked to socially directed or communicative behaviour in such a comprehensive way that, in the case of attempted suicide, there is not much value in using one to explain the other. Indeed, using the language of distress to explain a psychological epidemic of anything during the twentieth century begs more questions than it answers, given that distress is constituted at the heart of – and is a conceptual guarantor for – the new project of psychiatric epidemiology.

Notions of 'incidence' – how regularly this phenomenon occurs – are also important. For behaviour to be considered culturally sanctioned it must be widely, perhaps even self-evidently intelligible. That is, the meaning of attempted suicide must be obvious and agreed upon. Once this happens, it becomes just another meaningful action that humans might perform in relevant situations. A communicative overdose becomes a widely intelligible response to interpersonal difficulties. Thus another shift occurs, exceeding the situations described throughout this thesis, a shift where objects are produced and stabilised, through exclusions and emphases, in fields of enquiry made possible by various techniques and practices. When this 'information is general' (in Kessel's words), people might actually start doing it more often, feeding back further into the epidemic.

Conventional notions of incidence and epidemics need to be radically reconceptualised. The analysis of social phenomena such as this overdosing epidemic through body-counting and statistical compilation and computation are severely limited. Not only do these approaches run these two stages together, but this collapses the first 'technical' stage into the more simplistic second stage, where more people are able to start acting in newly established, resonant ways.

Concluding thoughts

The neologism 'parasuicide' is proposed in the 1969 letter by Kreitman, Philip, Greer and Bagley. The term is advanced on the basis that the phenomenon is current, important and generally established. In proposing the new term, it is noted that

> [t]he only point on which everyone seems to be agreed is that the existing term 'attempted suicide' is highly unsatisfactory, for the excellent reason that the great majority of patients so designated are not in fact attempting suicide.[139]

The neologism is also part of a local effort to refocus the Edinburgh Unit's energies, as it is soon to be explicitly reorganised around parasuicide (in 1971).[140] However, this local context should not obscure the more widespread agreement that a stable and distinctive pattern of behaviour exists. This pattern is based upon the newly self-evident fact that the great majority of attempted suicides are not read as having an uncomplicated intent to end their lives, but are in fact doing something else – something communicative and social. This chapter has shown how a particular vision of the social setting is constructed through a number of specific practices, ideas, assumptions and prejudices. The specifics of the 'social setting' should not obscure the principal point that people in the above studies, presenting at hospitals after having harmed themselves, are not being asked about their internal, emotional states at the time of the overdose, or about their family history of mental illness. They are being questioned about their social setting, their relationships with others, the people with whom they might be communicating – all this in order to make sense of the attempt. The idea of the significance of self-harm, an idea which seems so stable in the 1969 letter ('everyone seems to be agreed'), is to change radically over the next decade, as we shall see in the next chapter. The idea of self-cutting as tension-release is already being argued for by 1969, principally in North America. The link between Britain and North America is further strengthened as both countries loudly proclaim their affinity for neo-liberal economics in the 1980s. The links between the two countries in definitions of self-harming behaviour are also strong and increase in influence throughout the 1960s and 1970s. Underlying both neo-liberalism and self-cutting is a reading of human nature that is significantly more individuated and self-regulating than what came immediately before; social welfare and social communication give way (unevenly and gradually) to individuated emotional regulation, and eventually to biomedical, neurochemical ideas about self-harm.

Except where otherwise noted, this work is licensed under a Creative Commons Attribution 3.0 Unported License. To view a copy of this license, visit http://creativecommons.org/licenses/by/3.0/

5

Self-Harm as Self-Cutting: Inpatients and Internal Tension

At the start of the 1970s, the number of people recorded as 'self-poisoning as communication' is still rising. Typical is a 1972 report from Dunfermline that claims acute 'poisoning has reached epidemic proportions...[t]he number of poisoned patients increases year by year and there is no evidence that the trend is altering'.[1] In the same year, a bleak study issues from Sheffield, entitled 'Self-Poisoning with Drugs: A Worsening Situation'. This study claims that the rate of self-poisoning in Sheffield has doubled in the last decade and now accounts for almost one in ten medical admissions and one in five emergencies. Studies from Edinburgh, Oxford and Cardiff are cited as nationwide support for these truly alarming statistics.[2] By the late 1970s however, it is reported from the Edinburgh RPTC that rates of self-poisoning are falling for men and levelling off for women. Keith Hawton and colleagues in Oxford report five years later that overall 'the recent epidemic of deliberate self-poisoning may have reached a peak' around 1973.[3] Work on this phenomenon of self-poisoning, parasuicide or overdosing continues throughout the decade; clinicians marvel at the seemingly endless increase, and then wonder at the abrupt levelling-off. There are three major research centres for these studies: in Edinburgh, at the MRC Unit and Ward 3 of the Royal Infirmary of Edinburgh; in Bristol, at the Accident Emergency Department of the Bristol Royal Infirmary; and in Oxford at the John Radcliffe (General) Hospital. These endeavours are increasingly led by Norman Kreitman (Edinburgh), Hugh Gethin Morgan (Bristol) and Keith Hawton (Oxford).

Another form of self-harm emerges in the 1960s and 1970s in British psychiatry. Self-injury, self-mutilation or self-laceration are labels identifying people who damage themselves principally by cutting the skin on their forearms and/or wrists. This kind of self-harming behaviour is today the archetype broadly presumed to be indicated by the terms 'self-damage' or 'self-harm'. The rise in the prominence of this behaviour coincides with a decline in self-evidence for self-poisoning as communication, a cry for help. Overdosing comes to be seen (especially by those who focus predominantly on self-cutting) as an earnest attempt to end life, rather than a cry for help. This chapter brings into focus a clinical concern that, in a certain sense, displaces overdosing. This is not to comment upon the relative prevalence of these behaviours (a topic fraught with difficulty, especially around self-cutting), but to mark a transformation in what it meant by 'self-harm': from communicative overdosing to self-cutting performed for quite different reasons.

Like self-poisoning, self-cutting or self-mutilation does not have a common-sense, self-evident existence. It is a concept made and refined over a period of time, one which gradually becomes coherent and even obvious. What starts as a range of disruptive behaviours (including window-smashing, shouting obscenities, or swallowing 'bizarre' objects such as dominoes) is refined through increasing focus on self-cutting and the exclusion or relegation of other behaviours to secondary significance. Similarly, the reasoning put forth by psychiatrists in the earlier studies to explain the motivations for self-cutting oscillate between an awareness of communicative intent and a focus on internal emotional states that are regulated by cutting. Later on, this latter motivation becomes dominant. In these two ways, through practices of exclusion and emphasis, 'self-cutting as emotional regulation' becomes a coherent clinical concern, and it largely displaces the concern around self-poisoning. This move from socially embedded to internally self-regulating self-harm has particular salience given the political fracturing of consensus around welfare and the ascendancy of a neo-liberal rhetoric of self-reliance.

It is important to note that that clinical and psychiatric concern around self-damaging behaviour under the labels 'self-injury' or 'self-mutilation' existed in Victorian psychiatry, but did not refer to the kinds of self-cutting discussed here.[4] In fact, these terms have histories of their own, prior to the period covered here, and thus none of these terms should be seen as self-evident – instead, they make sense of particular behaviours in particular contexts. The clinical concept of self-cutting charted here is merely one particular way in which self-damaging behaviour is categorised. In the discussion of the various studies of self-cutting that follows, I have attempted to retain the terminology used by each author or group of authors, but this should not obscure their confidence that they are talking about the same phenomenon.

However, it would be misleading to say that cutting is entirely new in the context of self-poisoning or attempted-suicide studies: Batchelor and Napier, Stengel and Cook, and Kessel all report of people presenting at hospital having lacerated themselves. Sometimes this is implied by mention of surgical treatment;[5] at other times it is stated explicitly, as by Kessel in 1962, who notes that whilst gassing, throat- and wrist-cutting used to be common, but 'nowadays these come a poor second to drug taking'. Nevertheless, in Edinburgh's Ward 3 'patients with surgical emergencies resulting from attempted suicide – the cut throat and slashed wrists – are also managed in the ward'.[6]

Some general hospital-based studies during the 1970s use the term 'deliberate self-harm' to describe all methods of self-damage. Hugh Gethin Morgan claims in 1975 that this term is innovative, and he uses it because of his dissatisfaction with the other terms. Attempted suicide is said to imply that the intention is to commit suicide and, similarly, the term parasuicide 'might also be criticised for implying a resemblance to suicide'. It is further claimed:

> The use of 'deliberate self-injury' as a general term to cover the whole problem is itself ambiguous because it is often taken to refer only to physical injury, to the exclusion of drug overdosage or use of non ingestants.[7]

Morgan and colleagues thus use deliberate self-harm to cover overdose, non-ingestants and physical injury, including cutting. Even in the mid-1970s, Morgan and his collaborators are clearly concerned to include what they call 'laceration' in their analysis, as it is the second-most encountered method in their study (although admittedly it trails far behind drug overdoses, 91.8% at 4.8%. Despite these terminological discussions and the separation implied by using two terms – overdose and self-injury – in the mid-1970s general hospital-based studies lacerations are not seen as differently motivated behaviour. By the late 1970s this has become an issue in psychological, motivational terms.

In 1977 Norman Kreitman seems almost exasperated that self-injury cases are brought to a Regional Poisoning Treatment Centre: 'Despite its label, the centre also receives cases of self-injury presenting at the Royal Infirmary'. He reveals that one in 20 admissions to a poisoning treatment centre have injured themselves in ways other than poisoning.[8] However, as in Morgan's analysis, these cases are seen as merely methodological quirks. To be clear: these self-lacerators are a methodological minority, a small number of people whose supposed self-damaging communication happens to take a different form. There is no sense from these general hospital-based epidemiologists and clinicians that self-lacerators might be motivated differently to the self-poisoners.

The idea that this is a psychologically distinctive form of self-damaging behaviour emerges most prominently in North America. As Barbara Brickman and the present author have shown, a relatively coherent corpus of psychiatric journal articles emerges throughout the 1960s, with a particularly influential cluster published between 1967 and 1971.[9] These articles promote the view that behaviours called self-cutting, wrist-cutting, wrist-slashing, delicate cutting or self-mutilation exhibit 'much of the stability of a syndrome'.[10] These articles focus attention upon the behaviour of cutting the forearms or wrists and argue that it is predominantly found in young, physically attractive, intelligent female psychiatric inpatients. The cutting is said to be motivated by feelings of intolerable psychological tension, feelings that abate after cutting has been performed – often in a carefully considered and ritualistic manner.[11] These articles are at the root of the current clinical picture for what is today called 'Deliberate Self Harm' (DSH). Not only are the vast majority of these articles researched and written in North America, they are also predominantly from psychoanalytically influenced institutions, and all involve the study of psychiatric inpatients. This literature will not be re-examined here, as this would be largely repeating previous scholarship. However, the influence that this body of work has in Britain will be charted.

British literature on self-cutting in the 1960s and 70s is much scarcer, but that which exists is also overwhelmingly focused upon psychiatric inpatients. This is a key contrast to the self-poisoning studies which, as we have seen, focus upon people presenting at general hospitals' accident and emergency departments (These are also called 'community studies', as the people are not inpatients, but are living 'in the community'). 1960s–70s literature also contrasts with the current literature on self-cutting, which overwhelmingly focuses upon people who are not inpatients. Indeed, the concern with self-cutting in recent years casts it as an epidemic in the community, with the result that its emergence as a concern within psychiatric hospitals is rather obscured. This British literature forms the basis of this final chapter. In sum, this chapter seeks to investigate the emergence of a concept of self-cutting in Britain and how this meshes with the socially embedded attempted-suicide studies of self-poisoning that are overwhelmingly dominant in the British literature on self-harm until the late 1970s.

First, there is a brief restatement of the ways in which self-cutting and self-poisoning are differentiated in current clinical and counselling literature. Then we see how self-cutting emerges in Britain, with explicit influence from the American work. One aspect of the rise of self-cutting that has gone largely unremarked is that the behaviour first surfaces in the context of epidemic pathological behaviour – the spread of a behaviour pattern (self-cutting) in an institution, with focus upon how to control, manage, and eventually stop the spread of people performing the behaviour. As the 1960s progresses in Britain, this social-management approach gives way to a much more internally focused perspective, with emphasis on subjective feelings of tension and the falling away of imitative and communicative frames of reference. Today's model of self-cutting emerges as part of a move away from concerns about learning, contagion and imitation, and as part of an increased focus upon personality types, frustration thresholds and psychic tension. Once this inpatient phenomenon stabilises in the mid-1970s, it then informs the study of people who present at A&E departments, having cut themselves – a group briefly acknowledged but largely ignored in the context of self-poisoning studies. As noted, self-cutters at A&E are not initially perceived as psychologically distinct from the overwhelming majority of self-poisoners. This perception begins to change in the late 1970s. Finally, the reasons for the difference in inpatient and A&E objects of self-harm are briefly explored. Self-cutting behaviour seems to become the object of intensive psychiatric scrutiny relatively rarely outside of inpatient institutions (although it does register at A&E). Most individuals in these inpatient studies are admitted for other reasons, such as eating disorders or hysterical paresis. Initially, cutting only becomes scrutinised when inside the high-surveillance environment of a psychiatric inpatient ward.

Self-injury as self-cutting: the exclusion of overdoses in the present

The new DSM-5 category of non-suicidal self-injury (NSSI) excludes self-poisoning, which is described as 'intentional self-inflicted damage to the surface of his or her body'. With the specification of surface, self-poisoning is ruled out.[12] However, general hospitals still include both cutting and poisoning under 'self-harm' in their statistics. As seen in the Introduction, there is in the literature a strong differentiation of motives between cutting and overdosing – a differentiation that deals exclusively with self-cutting. These studies tend to be smaller scale, qualitative, and interview-based.

This differentiation between self-cutting and self-poisoning is varied and complex. It is largely achieved through four interlinked strategies, which can be labelled as: general assertion, motivational ambiguity, visibility and clinical management. General assertions are often rather sweeping statements, such as Tantam's and Huband's claim in 2009 regarding the 'very different cultural and psychological roots of self-injury and self-poisoning'. For them, self-injury means solely 'cutting, burning or otherwise damaging the skin and its underlying tissue'.[13]

In 2006, clinician Leonard Fagin begins with a general assertion, but then develops this into a comment on the motivations behind the behaviour:

> I see self-injury as different from self-poisoning, where substances (usually drugs) are ingested, usually in order to die, cry for help or obtain temporary respite from unhappiness or unbearable distress, and I believe that people who poison themselves have different characteristics from those who injure themselves.[14]

Note that the behaviours have been separated along with the motivations. Self-poisoners are still seen as crying for help, and the 'unbearable distress' is a rather precise echo of Kessel, but there is also a link with an earnest wish to kill oneself. With these conflicting possible motivations, self-poisoning is rendered ambiguous and unstable.

As far back as 1988, Barent Walsh's and Paul Rosen's book, *Self-Mutilation*, contains the following passage based on a criterion of visibility, and then develops into an argument about ambiguity of motivation. This passage is quoted by Armando Favazza in 2011 as 'the best explanation' for maintaining the difference between self-cutting and self-poisoning:

> In the case of ingesting pills or poison, the harm caused is uncertain, ambiguous, unpredictable, and basically invisible. In the case of self-laceration the degree of self-harm is clear, unambiguous, predictable as to course and highly visible. In addition, self-laceration often results in sustained or permanent visible disfigurements to the body, which is not the case with overdose. In various ways, therefore, these two forms of self-harm are quite different; the danger in combining them in a single category is that these important differences (and their clinical implications) are overlooked.[15]

In 2007 Jan Sutton differentiates between the behaviours along precisely these lines, arguing that self-poisoning is invisible and self-cutting visible, and therefore motivations for self-poisoning are ambiguous whereas for self-cutting, the intent is clear. She claims that 'self-injury is now well recognised as a coping mechanism and survival strategy, whereas the intent behind self-poisoning is less clear … It could be a botched suicide attempt, it could be an accident, it could be a cry for help, or it could be a means of temporarily escaping from emotional turmoil'.[16]

As well as visible versus invisible harm, and ambiguous versus clear motivation, the behaviours are further separated by clinical management strategies. In 2008 Pengelly et al. contribute to a debate about 'harm minimisation', building upon National Institute for Clinical Excellence (NICE) guidelines from 2004. Their guidelines include: 'If you feel you must cut, only use clean, sharp instruments to reduce the risk of infection and complications. Keep tetanus protection up-to-date … Avoid alcohol and drug use as you may inflict worse wounds than intended … Gradually reduce the severity of your injuries. Leave more time between injuries'. This practical minimisation attitude disappears when it comes to poisoning, as they state: 'Do not take tablets. There are no safe overdoses – even "small" overdoses can kill'.[17] Whilst self-cutting can be managed and minimised, self-poisoning must be prohibited. This feeds off and feeds into the stronger association with death that self-poisoning acquires between the late 1970s and the present. It is important to stress that I am not contesting any of this advice, merely pointing out that in terms of visibility, motivation and management, self-poisoning and self-cutting are strongly differentiated. All this effort confounds Favazza's assertion in 2011 that 'the British literature still does not make this distinction' between self-injury and overdosing.[18]

However, Favazza is partially correct – there is a British literature that persists in combining self-poisoning and self-cutting – primarily general hospital–based psychiatric epidemiology. These professionals largely conduct studies from accident and emergency departments as well as attempt to record the prevalence of self-harm that does not present to hospital but is established by retrospective questionnaire. A 2010 report by the Royal College of Psychiatrists states that '[f]or the purpose of this report we define self-harm as an intentional act of self-poisoning or self-injury irrespective of the type of motivation or degree of suicidal intent. Thus it includes suicide attempts as well as acts where little or no suicidal intent is involved (e.g., where people harm themselves to reduce internal tension, distract themselves from intolerable situations, as a form of interpersonal communication of distress or other difficult feelings, or to punish themselves).[19] To take another recent example, Hawton, Saunders and O'Connor define their object of study in a 2012 *Lancet* paper thus: 'Self-harm refers to intentional self-poisoning or self-injury, irrespective of type of motive or the extent of suicidal intent'. Self-poisoning and self-cutting are thereby combined. But it is not as simple as that, as they differentiate the behaviours in terms of incidence, claiming 'Self-cutting is the most common method of self-harm in adolescents in the community' whereas for 'adolescents presenting to hospital after self-harm...self-poisoning is by far the most common [method]'. They also differentiate by motive: '[I]ndividuals who self-harm by cutting differ somewhat from those who take overdoses, with suicidal intention more often indicated for self-poisoning, and self-punishment and tension relief for self-cutting'.[20] This is the same motivational differentiation shown above: suicidal intention against tension relief, which maps reliably onto self-poisoning against self-cutting.

However, this nuance in the epidemiological studies is not often reported by the literature focusing upon self-cutting alone, even though – despite some differences – they present extremely similar clinical pictures. Thus, Sutton is widely understood when complaining that hospital statistics under the term 'self-inflicted injuries' contain 90% overdoses: 'What sort of image does that [term] conjure up? Overdosing? I doubt it. Cutting? Highly probable...mention the word "self-harm", and it immediately conjures up images of people cutting themselves'.[21] Recent books on self-harm have titles like *The Tender Cut* (2011) and *Blades Blood and Bandages* (2012), and recent novels about self-injury are entitled *Cut* (2009) and *Scars* (2011), leaving little doubt about the methods of self-harm employed.[22] Whilst Sutton is right that there is a mismatch between the stereotypes that the term 'self-injury' conjures up (self-cutting), and the majority of people figuring in hospital statistics – 90% self-poisoning – this has not always been the case, as this book has shown in detail.

The scope of this chapter is not broad enough to focus upon all aspects of self-cutting, and instead focuses upon just one: the way in which self-cutting becomes conceptualised as a behaviour motivated by internal emotional states, rather than as a communication. This approach is in order to show how self-cutting becomes different from self-poisoning, which as we have seen, is intimately connected to communication and the social setting. As Shelly James points out in a recent dissertation, the reasons most often put forward for deliberate self-harm centre upon the relief of distress, a way of regulating or combatting emotional numbness. She also notes that social aspects remain under-explored.[23] James's dissertation may well be part of a swing back towards more socially embedded explanations but, if so, such a shift has yet to gather much pace or influence. The focus of this chapter, drawing upon the contrasts with self-poisoning, means that some important parts of the self-cutting stereotype are not addressed. The part of this focussing-down that has caused me the most disquiet is the lack of attention to some of the gendered aspects of self-cutting. The idea that it is an extension of grooming behaviour (said to be more prominent in the female psyche) or a practice rooted in vicarious menstruation, cannot be fully explored here. The analytical heart of this book is the epidemic of self-poisoning, not the practice of self-cutting, and difficult choices have to be made. This makes the following a rather partial and fragmented account of self-cutting, but hopefully a full and coherent account of the ways in which self-cutting and self-poisoning interact. As we saw in the introduction's analysis of successive editions of Myre Sim's textbook, some awareness of self-cutting, wrist-cutting or wrist-scratching, linked to affect regulation, emerges between the end of the 1960s and the mid-1970s, with a significant nod to North American clinicians. Cutting becomes archetypal in the 1980s, and as will be discussed in the conclusion, resonates with neurochemical explanations of human behaviour.

British clinicians and self-injury: inpatients and American influence

How does self-cutting or self-injury emerge in Britain? In what ways and through which channels does awareness crystallise and stabilise? Sarah Chaney has written of the various self-mutilating practices in Victorian literature and psychiatry, but at issue here is the specific phenomenon of self-cutting that emerges in the 1960s – something that Chaney acknowledges as rather different: 'self-cutting, often regarded a prevalent method of self-harm in the mid- to late-twentieth century, is not emphasised in nineteenth-century writings'.[24]

In British psychiatry, the story of self-cutting begins in Chicago. The principal study consistently referenced throughout the early British and American work on self-cutting is by Daniel Offer and Peter Barglow, psychiatrists at the Institute for Psychiatric Research and Training, which is commonly referred to by the acronym PPI. PPI is part of the private, Michael Reese (General) Hospital in Chicago, and in 1964 it is a 'psychiatric establishment [which] has a national reputation, especially for its research and teaching functions'.[25] Offer's and Barglow's study concerns an 'outbreak' of self-mutilation amongst adolescent and young adult inpatients over the nine months between November 1958 and August 1959, comprising 'approximately 90 incidents of self-mutilation'. Although they relate that '[i]solated incidents of self-mutilation had occurred periodically during the eight-year history of the institution',[26] the scale of this outbreak is unprecedented. PPI has a largely psychoanalytic or 'dynamic' approach, but Offer's and Barglow's conceptual approach

> follows the social-field multilevel approach illustrated by the hospital studies of Stanton and Schwartz. A field method was used because it became apparent early that self-mutilation was a complex product of many interacting and interdependent factors. Its ramifications extended throughout most of the hospital structure, and etiological factors could not be meaningfully evaluated in isolation.[27]

The approach of Alfred Stanton and Morris Schwartz (a psychiatrist and a sociologist, respectively), involves analysing the mental hospital in terms of relationships amongst staff members and between staff and patients, and of pathological symptoms (as far as possible) as social responses to conditions.[28] Intriguingly, a much bigger sociological study is being carried out at PPI at this time, led by Anselm Strauss, a pioneering medical sociologist who studies symbolic interactionism with Herbert Blumer and later associates with Howard S. Becker and Erving Goffman at the University of Chicago. In the book that emerges from this project, *Psychiatric Ideologies and Institutions* (1964), there is considerable analysis of what they call the 'Adolescent Scarification Crisis'. Again, this is tackled much less in terms of individual psychopathology and is far more about how institutions deal with crises. It contains large amounts of verbatim content from a conference hastily set up to deal with the fissures between staff members who become openly hostile to each other, arguing about the best way to deal with the 'scarification'. The sociological bent of Offer's and Barglow's psychiatric journal article coupled with the limited focus on individual symptomatology and pathology is striking testament to the influence of these sociologists.

Offer and Barglow still use the language of suicide to a significant extent, claiming that 'the self-mutilation incidents were "suicidal gestures" rather than "suicidal attempts"', where the latter signifies a genuine attempt to kill oneself. They argue that in all bar one incident,

'secondary gain' was involved, and some conscious effort to gain gratification from the environment was seen. Increased prestige in peer
group, desire for more attention from staff, competition with group
members, expression of anger toward family or hospital personnel,
were frequently encountered motives.

Only after this lengthy, socially focused list do they add their hypoth-
eses about how 'aggression and anger were then turned against the
self'.[29] What is important here is that a particular psychiatric symptom
(called scarification and self-mutilation) emerges from a psychiatric
inpatient facility in the course of sociological analysis. Nevertheless,
psychoanalysis can resonate with the social setting and interpersonal
relationships, through concepts such as transference (the transfer
of feelings from one social relationship to another, e.g., feelings for
a parent transferred to a therapist), or cathexis (the investment of
emotion into a person or object). The scarification is viewed in terms
of how it affects staff and staff relationships, its status as contagious,
and the roles of competition, bragging and attention-seeking that
might fuel it. It is an overwhelmingly socially embedded symptom,
with internal psychopathology subordinate to its social meaning and
social effects. Thus it has more in common with British communicative self-poisoning than with Asch's 'Wrist scratching as a symptom of
anhedonia' (1971), or with contemporary literature on self-cutting as
tension-regulation.

There is much British literature that focuses on the relationship
between inpatient institutions and psychopathology – for example,
Russell Barton's *Institutional Neurosis* (1959), and John Wing's and
George Brown's *Institutionalism and Schizophrenia* (1970). Illustratively,
the opening chapter of the latter is entitled 'Disease and the Social
Environment'. This period also sees the dawn of so-called anti-psychiatry, in which the sociological anthropology of Erving Goffman is so
influential. Indeed, as one historian expresses it, this is a time when 'the
diagnosis was social'.[30] As we have seen in previous chapters (especially
Chapter 2), this socially focused outlook has roots in the Second World
War. Tom Main, heavily involved in the second Northfield Experiment,
addresses the British Psychological Society in 1957 in what becomes one
of his best-known publications. Simply entitled 'The Ailment', Main
draws attention to the ways in which certain psychiatric inpatients
absorb disproportionate energy and attention from staff, creating problems, cliques and divisions within and between clinical staff.[31] Despite
this established seam of sociological influence, the inpatient literature

on self-cutting moves away from social explanations and – slowly and unevenly – emphasises internal psychopathology.

From imitation and epidemics to internal psychopathology and internal tension

In 1963, polymath and psychiatrist Colin McEvedy writes a dissertation on self-inflicted injuries for his diploma in psychological medicine (DPM) at the Institute of Psychiatry in South London. He joins the Maudsley in 1960 upon leaving his national service with the air force, and impresses the institute's director, Aubrey Lewis. McEvedy is best known for a controversial reinterpretation of 'Royal Free Disease', a 1955 epidemic among nurses at the Royal Free Hospital characterised by fatigue and ambiguous neurological signs. He and his co-author Bill Beard argue that this is a form of conversion hysteria, a conclusion that angers many.[32] McEvedy also publishes on hysterical epidemics in secondary schools: one of 'overbreathing' amongst schoolgirls in Blackburn, and one of vomiting, abdominal pain and 'faintness' in Portsmouth.[33] He is also well-known for his historical atlases. His analysis of self-inflicted injuries centres upon an outbreak of self-cutting in Bethlem and St Francis Hospitals, and a group of 13 patients in particular. I have been unable to obtain permission from McEvedy's next of kin to quote from this unpublished work, so I shall paraphrase throughout.

The only paper he finds that deals specifically with self-mutilation is Offer's and Barglow's (1960).[34] His work is partially concerned with the ways in which the behaviour might be learned or transmitted, but he also speculates upon the internal psychological reasons for behaviour that he considers to be bizarre and outside of recognised syndromes and symptom patterns. Crucially, of his opening case, Kay R., he relates that when discussing her behaviour with others he is questioned about the nature of her suicidal intent – using a continuum from a hysterical gesture designed to procure sympathy, to a 'genuine' attempt at self-killing. What strikes McEvedy is that he does not feel able to place Kay R. on this continuum. She does not seem to fit.[35]

Thus, McEvedy sees existing explanations as inadequate – this is neither an attempt at killing oneself, nor an attempt to elicit sympathy, nor a reaction to stress. The spectrum of possible action utilised by Stengel and Cook in the 1950s – between a social-stress reaction and a determined attempt to kill oneself – cannot accurately capture the actions of Kay R., the archetypal self-cutter in McEvedy's estimation.[36] He attends to the symbolism of these acts in a precise and sophisticated manner, arguing that even though popular opinion might hold that slashed wrists are lethal – and thus somebody might genuinely attempt to die by performing that action – he reasons that Kay R. must have realised quite swiftly thather wrist-cutting was more or less nonlethal, given that she kept surviving.[37] McEvedy's work shows how a certain form of self-cutting comes to light in Britain through its *separation* from socially embedded understandings of psychopathology – and this despite his more famous work on hysterical epidemics.

There are a number of intellectual assumptions that make this Bethlem and St Francis group a coherent set of 13 patients. The purported similarities between the cases exist alongside an awareness of the varied nature of their pathological behaviours. They are selected because they repeatedly self-lacerate, but also share many other characteristics. They are all young and female, and their psychological problems are seen to take many antisocial forms, including screaming, rudeness and obscenity, smashing windows and crockery, or swallowing unusual objects and taking overdoses.[38] Additionally, ten of the thirteen casesare thought to simulate illnesses or fits, to exhibit conversion symptoms, tohave fits that are not considered totally genuine as well as hallucinations thought hysterical rather than psychotic.[39] These patients selected for their cutting might manifest disturbance in very different ways, but these varied outcomes are thought to be rooted in (the same) impulsive paroxysm.

Despite this variation, it is self-injury that McEvedy investigates, and the patients all show injuries – some caused during aggressive outbursts (for example, window-smashing) and some cuts deliberately self-inflicted. He also mentions that a separate record is kept of overdoses and of any 'bizarre' swallowed objects. What is interesting, not to mention odd, to the sensibilities of the twenty-first century, is that injuries inflicted by window-smashing are included with self-cutting. This is perhaps because – as mentioned by clinicians below – window-smashing does not necessarily involve injury to oneself. However, in the controlled inpatient environment, it may appear as an obvious way to procure the sharp edges needed for self-cutting). Less jarring, but no less important, is the fact that overdoses are kept separate. This is one of the first examples that I have found of self-inflicted cutting (even though it includes window-smashing) being kept explicitly separate from overdosing. The claim is also made that self-laceration follows a remarkably consistent pattern, with the left wrist being cut most commonly (just less than half of all self-cutting incidents). Various behaviours are downplayed in order to cohere the group, as McEvedy refers to these patients – supposedly for reasons of brevity – as 'cutters'.[40] In this study the behaviour of 'cutters' is significantly differentiated from socially embedded hysterical, cry-for-help communications as well as earnest suicidal attempts. There is a sense in which this behaviour is new and unsettling (repeatedly labelled 'bizarre') and that it repeatedly (if subtly) confounds existing categories. This clinical object is not self-evident, but is the result of human analysis and intervention.

McEvedy's separation of this action from the social setting is based upon a differentiation between 'spontaneous' and 'susceptible' cutters – between those who perform the action on their own initiative, and those who try it in response to another patient's actions. There is a subtle relationship between these groups, but there is little doubt that the copying is – ultimately – secondary to the unity of the syndrome of cutting. McEvedy notes, with regret, that he would like to re-categorise the patients in order to separate out those who perform the act without any imitation of others. However, isolating these non-imitators requires too much of a reworking of the material. He has no doubt that some of his Bethlem group are only classed as repeated self-lacerators (those with five or more cutting incidents) because they happen to be present during the self-cutting epidemic.[41] This makes it clear that the key to the syndrome is in the internal impulse, not the social imitation.

McEvedy argues that the most notable aspect of behaviour is the apparently unprovoked mood swings – bringing emotional states to the fore. These are distanced from the social environment in the case of Kay R., who is said to have cut herself over and over, regardless of environment or levels of stress. McEvedy finally distances cutting from a socially motivated phenomenon because the impulse (presumed to underlie all the behaviours – from cutting forearms to swallowing dominoes) seems so unorthodox that it cannot be explained by mere social pressure or stress. He reasons that there must be something preventing the (supposedly suicidal) impulse from being conventionally expressed.[42] That it might also be connected with hysterical, susceptible imitators does not change the fact that the behaviour begins in a pathological, emotional, internal impulse rather than a disordered social setting.

McEvedy is not entirely sure what might replace the powerful aetiological force of the social environment or imitation, but he speculates thatthe so-called 'spontaneous' cutter's personality has not only a high level of hysterical traits, but also something that he labels 'hostile tension'.[43] This is the first mention of 'tension' as a key motive force for 'cutters' in Britain – something that is well established in the

contemporary literature, and it emerges in McEvedy's work as characteristic of the cutter who is not simply copying.[44]

In sum, McEvedy makes a particular effort to distance the patients
from socially embedded motivations. The research is not written up
into any research articles and thus does not garner much attention or
influence, although it is available at the library of the most influential
psychiatric training institution in Britain, the IoP. Unlike his work on
the 'Royal Free Epidemic', which is explained as conversion hysteria,
this study remains largely obscure.[45] It is referenced by a study from a
Plymouth adolescent unit in 1968, a more influential unpublished study
in 1972, and a published article in the *British Journal of Psychiatry* in
1975 (all of which are considered below). Its significance lies in the fact
that it isolates a group of female psychiatric inpatients, focuses upon
one particular symptom, and presents in such a way that the impulse
underlying that particular symptom is important, rather than its status
as an epidemic behaviour or shedding light upon the social organisation
of a hospital. Although bizarre and unorthodox to McEvedy, self-cutting
in response to 'hostile tension' seems very familiar to us.[46]

Later in the 1960s, D.W. McKerracher, a clinical psychologist at
Rampton secure hospital in Nottinghamshire, publishes two articles of
note with a number of different colleagues, all working at the hospital.
There is an established literature on prison self-mutilation that emphasises self-harm as a response to the confined space, or to perceived
injustices.[47] A secure hospital environment can feed off that frame of
reference, but its status as a secure psychiatric hospital means that staff
are likely to give close consideration to internal and psychopathological
factors. In 1966 a comparison of the behavioural problems of male and
female prisoners in the hospital is published. In 1968 there emerges a
specific study of self-mutilation in 'female psychopaths'.[48] These articles are important because they show how self-mutilation becomes
more strongly established through understandings of internal psycho-
pathology.The behaviour is seen as less outward-looking, social and
communicative in its meaning, and more internal and emotional. It is
seen as a ritualistic behaviour predominantly performed by females, and
it clearly troubles the clinicians. However, it is also grouped together (in
the second study) with the practice of 'window smashing'. Again, the
familiarity of some of the observations jars with this detail. This marks
it out, as with McEvedy's study, as an inpatient phenomenon, and we
have glimpsed it in McEvedy's study, too. The significance of window-
smashing is difficult to ascertain: it is often seen as merely an expression of vandalism and also as connected to experiences of confinement.

A psychologist at a Durham remand centre in 1981 conducts a study on the smashing of cell windows. He observes that 'window smashing is predominantly "expressive behaviour", stemming from boredom and frustration'.[49] This obviously does not exhaust the significance of this action but it is relevant that this problem (for McEvedy as well as McKerracher and colleagues) is predominantly being studied in people who are confined.

The first study, by McKerracher, Street and Segal, is a general comparison of behavioural problems, with focus on aggression in particular. The appearance of general, non-specific aggressive behaviour in women is seen as less understandable than is that same behaviour in men. They comment that 'men seldom seemed to indulge in aggression merely for the sense of release obtained from it': by implication, women did. The aggressive outbursts are tentatively characterised as 'displacement activity which helps patients to avoid experiencing feelings of anxiety and subjective stress'. They argue explicitly that the 'aggression of the females, however, seems more emotional than instrumental, and erupts spontaneously whenever they feel angry, tense, anxious or even depressed'. The most striking formulation for contemporary accounts of self-cutting, however, is the following: 'They seem to experience feelings of internal stress which build up to such a state of tension that violent activity becomes essential'.[50] However, it must be born in mind that they are talking about all female aggression – to property, to themselves, to staff, verbal threats, threatening suicide or even refusing food. A whole host of behaviours can be reduced to these emotional outbursts. The aggression is theorised in terms largely independent of the confined surroundings, with explanations focused upon 'a stronger primary drive level of anger' and 'lower frustration thresholds'.[51]

With different colleagues, McKerracher publishes specifically upon 'self-mutilation', focusing upon a group of 'female psychopaths'.[52] The authors refer briefly to the results of the previous study with the claim that 'female patients were significantly more prone than males to mutilate their own bodies and smash hospital property'. They note that these incidents are normally regarded as 'hysterical', and they compare two groups: one that 'indulged in self-mutilation and smashing of windows' and another, slightly smaller group, that does neither. The authors expand upon this relationship between self-mutilation and window-smashing, observing that '[m]any of them had smashed windows for the purpose of self-mutilation though this should not be taken to imply that all window-smashers are necessarily self-mutilators'. They quote a personal communication from a colleague who claims that

> in some patients window-smashing is indeed a means of self-mutilation, and in fact these patients often adopt other means to the same end; but there are other patients who regularly smash windows in a way that causes them no injury at all, and their actions can only be regarded as aggressive towards others and not towards themselves.[53]

Thus the behaviours of self-mutilation and window-smashing are combined for the study, but McKerracher and colleagues are explicitly aware that they can – and perhaps should – be differentiated. This division along lines of inward- and outward-facing aggression seems to herald a weakening of the association between the two behaviours.

The Rampton clinicians read Offer's and Barglow's 'important' study as claiming that 'attention-seeking, prestige-gaining and tension reduction were the main goals of self-mutilation', and that 'the major dynamic was aggression turned against the self'. As we have seen above, the study from PPI in Chicago is significantly more focused upon the former: the social, institutional and epidemic aspects of the behaviour.[54] What is striking about the Rampton study of incarcerated patients, 'who could loosely be termed feeble-minded psychopaths', is that there are a number of links with current literature, specifically on role of the cutting as ritualistic behaviour and in reducing internal psychic tension.

It is hypothesised that 'the acts of self-mutilation and window-smashing may have a ceremonial or ritual quality' that is made habitual by the positive reinforcement of 'tension reduction'.[55] It is important to clarify the difference between the anthropologically influenced ideas of ritual cutting practices (such as penile subcision) which are often excluded from contemporary ideas of self-cutting, and a more general description of ritualistic practices, which suggest the establishment of an informal but highly habituated set of actions that a person might perform before and after carrying out the act. It is notable that ritual and tension reduction (prominent in the current literature on self-cutting) are here assumed to play a role that underlies both self-mutilation and window-smashing.

This focus upon individual reinforcement due to tension reduction is a less socially focused way of explaining the behaviour than Offer and Barglow, for example, but this is not the whole story. The Rampton clinicians argue that the supposedly 'horrifying form' of the 'compulsive "acting-out"' is linked to the 'restrictions of a security environment' and the 'limited range of activity available'. This feeds the '[s]uppressed interpersonal aggression occurring in a personality that has low thresholds of boredom and feelings of frustration'. The social environment is considered very important here, but in a way that bears much more explicitly upon individual, psychological needs – unlike McEvedy's analysis, where the social setting functions more to explain the transmission and imitation of the behaviour. McKerracher and colleagues class the mutilators and window-smashers as more 'obsessive-compulsive, phobic, and pre-occupied with bodily complaints'. There is no significant engagement with the sociological, epidemiological side of the preceding literature. As noted above, this is part of a more general shift towards personality types, frustration thresholds and psychic tension, and a step away from learning, contagion and imitation. This may be linked to their finding that '[s]urprisingly, hysteria was not a discriminating characteristic',[56] but whatever the cause, it seems highly significant in retrospect.

The same year, a study on epidemic self-injury by P.C. Matthews is published from an adolescent unit in Plymouth. The focus here is upon the spread and control of the behaviour, and includes a 'sociogram' that plots so-called 'ratings of social power' between the adolescents to try and make sense of the spread of the behaviour. The inner feelings of these patients are mentioned, but not analysed in any significant sense, as they do not seem to cohere in any logical manner.[57] The article does not give much space to the inward-looking, inner tension, psychopathology stance – it remains much more focused upon the epidemic, contagious nature of the symptoms, using self-mutilation to shed light on other potentially transferrable or imitative behaviours.The focus is upon management strategies to stop behaviour spreading rather than on investigation into the significance of the mutilation itself. So we can see that the shift described in this chapter is not a straightforward chronological progression from social fields to internal tension, but is a partial and uneven shift. However, as we go from the 1960s into the 1970s, this shift becomes increasingly apparent.

Presuming the social and confounding expectations

In 1970, an article is published from the Maudsley Hospital by J.P. Watson, one which gives us some insight into a doctor's expectations about self-mutilation motives and how these might be confounded or modified by clinical experience. Crucially for the shift being described here, the expectations concern the social setting, and the clinical results privilege internal emotional states. Watson describes one patient (as opposed to a group) and focuses upon the relationship between the patient and himself (her doctor). The article's central concern is why the patient might cut herself, and what various interpersonal relationships might have to do with it. In order to measure these relationships, Watson uses a 'repertory grid technique', a formalised way of processing interpersonal data regarding social roles and relationships. It is based upon American psychologist George Kelly's Personal Construct Theory. Various factors or relationships are rated a number of times by the patient according to their strength or significance in the individual's world-view (or personal construct).[58]

The patient is admitted to an unnamed psychiatric unit because of reported anxiety and depression that leads to her cutting herself 'with glass or razor blades on her arms and face'. This very visible behaviour is initially ascribed to 'difficulties with a boy-friend, G., of whom her parents disapproved'. Watson expects this social frame of reference to be an adequate explanation: 'When I began psychotherapy I thought that disturbed relationships with both parents and the unhappy experience with G. were probably the most important determinants of the patient's self-mutilant behaviour'.[59] However, according to the repertory grid,

> the elements 'having the same thoughts in my head for a long time', and 'wanting to talk to someone and being unable to', not the elements concerned with persons, were the situations ranked as most likely to make her cut herself, feel angry and depressed, and think people were unfriendly.[60]

Watson's expectations shift from social circumstances and interpersonal relationships, to internal thoughts and desires.

Although G. does feature rather more significantly than Watson expects, Watson's own presence, and that of her parents are not reported to be significant. Watson initially suggests that this may have to do with denial – given that the grid is a self-report technique – then concludes that this is 'a complex matter, but I think it likely that the "person" elements seemed to her less likely to upset her and make her cut herself than the "talking" and "thoughts" elements'.[61] Even though the article is based upon an individual, there is still much scope for relational, interpersonal aetiology (along the lines of self-poisoning explanations). However, this is rejected, seemingly on the basis of the patient's own reported statements. This shift in explanation becomes more and more established as the 1970s progresses.

In the same year, a dissertation for an MSc in clinical psychology at the IoP, entitled 'Self-Mutilation', is completed by psychologist Anthea Keller. Self-cutting is seen to have two possible causes, which possibly interact – internal and the external. Keller recognises that the under-standing of self-cutting as internally focused tension relief is prominent in the literature, and she references three articles from North America published in the late 1960s. All these references feature relatively regu-larly in the current literature on self-cutting, at least until the early years of the twenty-first century.[62] This might also be seen as the beginning of the explicit (referenced) influence of the North American psycho-analytic, internalist studies of the late 1960s on studies carried out in Britain – with previous influences limited to the sociologically minded Offer and Barglow.

The role of social or outward-looking factors is seen to be more diffi-cult to isolate or pin down with any precision, according to Keller, due in part to difficulties patients have in talking about it. In a similar way, Keller divides incidents into two classes: 'group' and 'individual' cuttings. The former concerns the sociologically influenced literature considered above, such as the studies at PPI by Offer and Barglow and Strauss et al. Keller claims that, apart from a higher proportion of men in some studies, there are no substantial differences between the two groups. She then argues that any 'group' cuttings only occur when individuals already have a predisposition to the behaviour.[63] In this way, under the veneer of parity, the individual cuttings are in fact made more significant, being the root cause of any group cuttings that may occur. This echoes McEvedy's rooting of the behaviour in 'sponta-neous' cutters who then influence 'susceptible' ones. For reasons both practical and theoretical, the group is secondary here to the individual inclination.

Despite this, some recognisably 'social' or relational factors are broached. Keller mentions that visits of parents and setbacks during therapy have been seen as significant in the aetiology of cutting. However, it is also claimed that virtually every published investigator of self-mutilation emphasises the role of building tension (which may not have any obvious reason behind it) that then overwhelms patients and causes them to try to reduce the tension by self-cutting or smashing windows.[64] Patients might cut when alone or feeling lonely, but also, confusingly, when in the presence of an important person. It is unclear to Keller why cutting happens in the latter scenario, given that soli-tary, affective relief is the dominant explanatory frame here. Window-smashing has not entirely retreated from consideration, but it is clear that the sociological, group-epidemic focus is fading, being replaced by a model of internal affective regulation – something that corresponds quite closely to today's understandings of self-injury.

Brian Ballinger publishes a study on self-mutilation in 1971, comparing two populations from Dundee: one group from Strathmartine Hospital for the 'mentally subnormal', and the Royal Dundee Liff Hospital, a psychiatric inpatient institution. Right away, Ballinger makes clear that he is not talking about window-breaking, restricting the study to acts that are 'painful or destructive...committed by the patient against his own body', and excluding'[a]ccidents, tearing clothes, window-breaking, swallowing dirt and refusal of food'.[65] Frustrated outbursts are here only included if they involve damage or pain (in the assessment of the staff) to the patients' bodies. There is a sizeable literature, which has been mentioned, around repetitive self-damage performed by those with severe learning difficulties; Ballinger's study is an explicit attempt to compare two recognised categories of self-injury: that of the 'mentally subnormal' and that of the 'mentally ill'. These categories remain very separate today, with little attention on the former.

The methods of injury are seen to differ, but with significant overlap. Patients in the 'subnormality hospital' are reported to self-injure by 'picking, striking, scratching, banging, biting, pulling hair out and rubbing'. The psychiatric patients, on the other hand, injure themselves by 'scratching, picking, striking, rubbing, cutting and tying string round fingers'. There is no sense here that self-cutting is an archetypal form of injury. Self-injury is seen as more prevalent in subnormality hospitals than psychiatric hospitals, with 15% of patients in the former institution engaging in self-injury, compared with only 3% of the latter. It is seen as related to the social setting: 'environmental restriction, boredom and frustration played a part in worsening self-injury in many patients'.[66] Here again it is not a smooth (teleological) progression from social explanations to internal ones, but this article is useful in showing how the concept of self-injury encompasses a number of distinct, but overlapping, inpatient populations, and becomes ever more visible – and differentiated – throughout the 1970s. Psychiatric self-injury is a clear, definable object here.

North American influence and the triumph of internal tension

One of the MPhil dissertations submitted at Institute of Psychiatry in 1972 is entitled 'Wrist-cutting: a Psychiatric Enquiry'. Little is known of the author, Samuel Stuart Anthony Waldenberg and, as far as I can make out, the research does not form the basis for research articles in psychiatric or medical journals. The study sits on the 'thesis' shelves in a newly refurbished section of the IoP, amongst the dissertations of some of the luminaries of British psychiatry, including Sidney Crown, Michael Rutter and Murray Parkes, as well as Neil Kessel and Norman Kreitman (and Colin McEvedy). Despite Waldenberg's relative obscurity, the study is referenced by a number of subsequent published texts on self-cutting.[67] The dissertation contains a sheet showing when it is signed out of the IoP, and by whom. Names on the sheet include Alec Roy, who publishes on self-mutilation later in the 1970s (see below), and then on depression, suicide and schizophrenia, and Dinesh Bhugra, currently professor of mental health and diversity at the IoP, and author of – among many other books – the Maudsley Monograph, *Culture and Self Harm: Attempted Suicide in South Asians in London* (1994).[68] As with McEvedy's work, I have been unable to obtain the necessary permissions from Waldenberg to quote from this unpublished thesis, so I shall paraphrase throughout.

Waldenberg's method is similar to McEvedy's, built around an 'accident book' at the Joint Royal Bethlem and Maudsey Hospitals, in which a record is kept of all the injuries to inpatients that come to the attention of the staff. He notes that a similar book exists at St Francis Hospital.[69] From these two sources, a group of self-injuring patients is selected for study, with patients interviewed soon after the incident. Their responses to this semi-structured interview are compared with a control group of non-cutting inpatients.

There is an effort to emphasise the cutting over and above a constellation of symptoms. He lists various behaviours occurring in this sample, including truanting, delinquency, the taking of illicit drugs, and supposed sexual deviance: lesbianism, promiscuity and incest. As for more directly and physically harming behaviours, he notes that these patients take overdoses (mostly with no suicidal intent) and engage in self-cutting, window-smashing, self-burning and self-scalding.[70] It is notable, given the content of previous chapters, that these patients often take overdoses without suicidal intent, but these are not investigated (a point developed below). Window-smashing is still considered an issue (although admittedly minor), as are other supposedly deviant behaviours. It is important that self-cutting is not self-evidently or obviously the behaviour at the centre of these patients' pathologies: it is made central by the emphases of professional observers. (The same processes of exclusion and emphasis operate in the North American literature.[71])

The key finding of his study, according to Waldenberg, is that internal, emotional gain experienced as a result of cutting is seen to trump any

kind of external, social gratification.[72] Explicitly then, this dissertation constitutes an argument against the socially embedded self-harm
analysed throughout this book. It is influenced by a number of North
American studies on self-cutting that promote internal, psychological,
emotional needs as the roots of self-cutting, especially tension release –
studies that continue to influence current models of self-harm. He refers
many times to North American studies by Pao (1969), Crabtree (1967),
Graff and Mallin (1967) and Grunebaum and Klerman (1969). Indeed,
he also mentions Offer's and Barglow's sociologically influenced study
of 1960 which, he admits, focuses more upon the role of imitation in
epidemics of self-cutting. He concedes that psychoanalytic authors such
as Pao, Crabtree and Graff and Mallin do not give the role of imitation
muchconsideration.[73] Imitation implies a social field, and the idea that
the point (underdeveloped in the American psychoanalytic studies)
shows how a division is opening up between the internal and external
ideas of causation. This division has the potential to separate any self-
cutting that might present at A&E departments from the overwhelming
mass of socially embedded and understood self-poisoners with whom
they are combined in the 1970s analyses of those such as Hugh Gethin
Morgan. However, Waldenberg does not make this split according
to method. He calls the group 'cutters' but argues that some of these
patients can distinguish between the feelings that precede a frankly
suicidal overdose and those that precede cutting and/or a less serious
overdose.[74] This equates cutting and trivial overdosing and implies that
they are prompted by the same state of mind. Thus, the strong differentiation between cutting and overdosing does not seem to stem from
here. However, the emphasis on cutting and on internal motivations –
explicitly against sociological or epidemic ones – is highly significant.
Self-cutting is cast as internally rather than externally motivated, but
this internal motivation is also ascribed to trivial overdoses.

Ping-Nie Pao's study of 'delicate cutters' from Chestnut Lodge,
Maryland, is praised for the clarity of its descriptions, especially the
patients' subjective experience of cutting. Waldenberg quotes Pao's
account, which uses the words *tense, tension,* and *tenseness* in a single
sentence. As well as this internal emotional state, Waldenberg does
acknowledge the social setting, mentioning interruptions in interpersonal relationships as possible factors that might precipitate cutting.[75]
His literature review is ambivalent about the internal/external divide.
He writes that others have noted the relationship between an episode of
cutting and interpersonal disturbances, such as the end of visits; others'
works might start with a view of cutting as a purely internally focused

activity, but then come to see it as a communication between patient and therapist.

These social motivations exist in tandem with acknowledgement of patients experiencing a painful sense of unreality, or of having no feelings at all, which prompt the patient to cut to try and relieve them. Similarly, the review mentions the dual desire on the part of the patient to punish the parents, but also to obtain their help and support (thereby to communicate with them), alongside rather frank statements that cutting is performed in order to relieve tension and to alleviate feelings of numbness or deadness.[76]

This initial ambiguity about the internal nature of self-injury is disciplined by copious and repetitive intellectual labour (of which Waldenberg is significantly aware) so that the clinical data conforms to his expectations. It is the clearest sign yet that a battle is being fought to de-couple self-cutting from socially focused, communicative action. This is partially fuelled by the belief of the psychiatrists (against other medical and nursing staff) that the behaviour is meaningful above and beyond simple attention-seeking or 'acting out'. Waldenberg notes that the staff (both nurses and doctors) often react negatively to these patients and label them as manipulative or attention-seeking. He downplays this angle, reasoning that there are numerous ways of seeking attention, yet these patients choose a method that – to him – is extremely unusual, even bizarre.[77] This particular point is not an aetiological argument formed from psychoanalytical inclinations (like much of the American literature): it has a much more mundane, everyday conflict at its heart – a conflict between those who see some psychiatric patients as manipulative (and therefore communicative) timewasters, and those who see another order of significance in their behaviour. He argues that, because most patients who cut do so whilst alone, the cutting therefore must serve internal needs, rather than communicative ones. This is a clear intervention against certain reactions to the behaviour and also seems to preclude – for Waldenberg– any attempt to link the behaviour to communicative overdoses. However, his reasoning does not quite hold, as communicative overdoses would also – in the majority of cases – be performed alone and later discovered, much like the cutting incidents. However, the internal tension-fuelled motivation becomes a powerful counter of legitimacy for the discrete nature of the behaviour pattern, as well as countering perceived negativity from other staff members. Indeed, when recounting all the other deviant behaviours, from truancy to incest to overdosing, Waldenberg relates the difficulty he has in isolating a single psychological motive or explanation for these various behaviours.[78] Thus he chooses to narrow the focus to explaining just one – the self-cutting – rather than taking a more general sociological, deviance-based approach that might attempt to make sense of all the behaviours as a group.

He acknowledges that patients vary in the reasons they give for their cutting, but that they normally allude to tension in some form. He further notes that despite the varying responses, tension and anger (both with oneself and others) feature prominently.[79] There is some ambivalence here: he acknowledges variation in explanations, and that interpersonal anger is a factor. He continually oscillates between awareness of social factors and emphasis on internal ones; he consistently promotes the internal rather than interpersonal causes. He addresses the issue of communication directly, acknowledging that even though any intent to send a message might be denied by patients, they also have obvious expectations about the reaction of staff to their behaviour. However, his final judgement is stubbornly internally focused. He claims that even those with such 'social' expectations report relief and satisfaction from the sight of their own blood, the experience of which outstrips any pleasure they might get from a doctor's reaction.[80]

Waldenberg discusses the views that his control group of 'non-cutters' have on the subject of cutting, which is reported as grudging approval at the discipline or 'nerve' required to cut oneself. This approval is presented by Waldenberg as evidence, but not for the way in which cutting becomes a socially acceptable, valorised and aspirational pattern of behaviour. Instead, it is deployed as evidence of the internal needs serviced by cutting – the control patients are presumed to have slightly less-powerful urges. He also claims that pleasure from bleeding is a 'simpler' explanation than anticipation of the therapist's reaction – a clearly loaded assessment.[81]

He does not deny that his group of 'wrist-cutters' receive much attention after cutting, and that they may indeed derive satisfaction from this attention, but he calls the gains from the act of cutting, itself, as 'primary', and that they outweigh the secondary, interpersonal effects. These primary gains are internal, emotional, and heavily psychoanalytic.[82] The motives of these patients are clearly multifaceted (and this is acknowledged), but the consistent emphasis is on the internal, affective regulation of cutting or of seeing blood. The number of times this oscillation is played out indicates how hard Waldenberg has to push against the socially embedded analyses of self-harm, especially in institutions. There is considerable room for a mixture of both causes, but one is emphasised, and he reproduces the conclusions of the influential American literature.

On a practical, methodological level, Waldenberg is in part aware of how the research methods employed might influence the findings in this way. He explicitly admits that whilst the patients' feelings and thoughts that immediately precede the cutting episode are subject to close examination and questioning, less attention is afforded to possible motivating factors in the patient's social circle. Such methodological candour and awareness is striking and shows how the focus of the dissertation, influenced by the North American literature and the local staff conflicts, emphasises the internal, psychic motivations over the social setting. It is acknowledged that an interviewer's questions can be 'leading' and that the information *sought* by the questioner is often furnished by the interviewee.[83] However, at the end of all these oscillations, Waldenberg's judgement call specifically emphasises the internal over the external.

Angela and Alan Gardner publish a study in 1975 from Long Grove Hospital in Epsom and the London Hospital in Whitechapel. They investigate a group of 22 female inpatients (8 from a psychiatric ward of the London Hospital, and 14 from Long Grove, a traditional mental hospital), who are admitted over the course of one year, from July 1972. They use the Middlesex Hospital Questionnaire and the Obsessive-Compulsive section of the Tavistock Inventory: both are psychiatric rating scales. The former is developed by Arthur Crisp and Sidney Crown during the mid-1960s as a rapid, self-report diagnostic tool for neurotic patients. The latter has its roots in the psychoanalytically oriented Tavistock Clinic.[84] These 22 patients are compared with a control group.

Gardner and Gardner claim that self-mutilation has been around for centuries, but only recently brought into focus by Offer and Barglow, whose work has 'focussed interest on the patients, usually female, who repeatedly cut their wrists'. They acknowledge that 'since then 'a number of reports have appeared, mainly from the U.S.A' and they are rather dismissive of British literature, stating that there have been '[o]nly three studies of consequence': McEvedy (1963), McKerracher et al. (1967) and Waldenberg (1972).[85] The North American provenance of most of the analysis of self-cutting is again implied.

Gardner and Gardner argue that both Offer's and Barglow's (1960) and McEvedy's (1963) studies show that 'repeated self-cutting appears to have an "infectious" quality, leading to outbreaks involving several patients. This suggests that factors in the ward milieu play their part'. They address this by selecting matched controls from the same wards as the cutters. However, they state explicitly that when interviewing the patients, '[s]pecial attention was paid to the patient's mental state during self-cutting'. This again shows the emphasis on an internal, psychological perspective over a socially focused enquiry. They report that the 'initially private nature of the act is well emphasized', which contains both an emphasis on being alone, but a concurrent awareness that the consequences of the cutting may be displayed later. Key is their contention that '[b]y far the commonest experience leading to self-cutting was the onset of an unpleasant feeling of tension, this increased in intensity until the patient cut her skin, which brought an immediate lessening of tension and a feeling of relief'. It is noted that this might have to do with the patient's social circle, but this is downplayed: 'Sometimes the feeling of tension was related to angry feelings towards self or others, but more often than not there was no apparent precipitating factor'.[86]

Despite this partial acknowledgement of the social setting, their entire therapeutic strategy is based around feelings of tension. They claim that for any treatment that attempts to halt self-cutting behaviour, it is 'logical to seek some other superior tension-relieving reward'. Gardner and Gardner do acknowledge that tension relief and communication might exist in the same action, as they characterise another article's findings 'regard[ing] the self-cutters' method of tension relief as a pre-verbal message'. They fail to establish any secure differences between cutters and controls, but claim that 'it remains possible, even probable, that differences do exist but are found perhaps in the quality of child/parent relationships and other areas difficult to assess with certainty in retrospect'.[87] Thus, they remain committed to the psychological discreteness of this population of 'self-mutilators'. This article again attempts to differentiate self-cutters from other kinds of psychiatric inpatient, on psychological grounds that are increasingly tension-focused.

Stability and comparison with self-poisoning

One of the final steps in this process that isolates 'self-cutters' as a distinctive object of psychiatric research and treatment in Britain (especially as the study of self-poisoning is so well established) is to compare these 'self-cutters' explicitly with a population of self-poisoners. This is done in 1975 by Michael A. Simpson, a clinician who publishes on the topic of medical education and later on borderline personality disorder (BPD), and who trains at Guy's Hospital in London. He conducts an interview survey on 24 self-cutting patients brought to his attention by a 'Psychiatric Emergency Services Unit, dealing with all requests for a psychiatric opinion from the general medical and surgical wards and also a busy Emergency Room' based at Guy's. He also produces the first comprehensive literature review on self-mutilation, which published the following year in a collection edited by the eminent North American suicidologist Edwin Schneidman.[88]

Simpson estimates the lethality of all the self-damaging acts he includes in his clinical study and concludes that all wrist-cutters fall into the bracket of lowest lethality, whilst the self-poisoners are more variable. What is striking about Simpson's ratings – at least for those familiar with the studies of self-poisoning produced during the 1960s and early 1970s – is that Simpson's ratings do not appear to include any assessment of the social setting. In contrast to Stengel's assessments (Chapter 2), which adjusted lethality ratings according to precautions taken to avoid or ensure discovery, Simpson's focus is decidedly on the mental state and experience of the patient, rather than on the social environment. As Simpson puts it, patients were 'interviewed with special reference to the phenomenology of the act of cutting'. (In this sense, 'phenomenology' indicates a focus on the subjective experience.) Accordingly, the entire enterprise is based upon patients' self-report of their feelings and motivations. They are 'asked to state their first and second most serious or troubling symptoms. Of the cutters, nine complained of depression as the first or second most serious symptom, twenty-one complained predominantly of "emptiness", and eighteen of tension'.[89] These statements match up with some of the previous studies, but what is most interesting here is how far the kinds of questions asked correspond to the quality of the answers. Simpson makes no mention of the social setting, patient relationships, or possible communication.

Thus, he reports that '[t]he non-cutters complained primarily of depression, each included it as one of the two principal symptoms, and nine cited tension as their second most-troublesome complaint'. It is important to remember that the people designated as 'non-cutters' are in fact self-poisoning patients. Simpson has little time for psychiatric diagnoses, arguing that it is 'not helpful with regard to wrist-cutters and they are best regarded as a separate category in planning management'. This shows the (still relevant) ancestry of the behaviour pattern as a sociologically influenced management problem. He reports: 'Nine of the present series of cutters absconded from hospital on numerous occasions, a pattern of behaviour which was not seen in any of the non-cutters'.[90] Again, this is a management issue after patients are admitted to psychiatric wards. As much as he focuses upon the subjective experience of the patients, there are many echoes of the social field and of sociological studies that focus upon the particularly intractable management problems presented by these patients, problems that are increasingly reduced to 'cutting'.

Despite these management issues, Simpson becomes increasingly confident that the behaviour is essentially a response to tension, even if also learned or contagious. He argues that '[t]his form of a response to tension can be learned and propagated in a hospital or institution and is often sustained by the widespread conflict and guilt such acts tend to arouse in the staff'. Any focus on the social or administrative setting is secondary to the essence of the behaviour, which is characterised as a 'response to tension'. The focus on internal, emotional states is combined with a desire for solitude, leaving any wider social or communicative significance out of the reckoning: 'The patient feels depressed, angry and tense, and wants to express the extent of her feelings, but feels unable to do so in words. Tension becomes the predominant affect ... she will seek solitude if she is not already alone'.[91] Simpson's study illustrates the lack of concern for the social setting with his assessments of lethality, the focus upon the subjective experience (phenomenology) of the cutting, and it shows the management issues that persist in assessments of self-cutting. Again, the focus is upon individual, intolerable tension as motivating the cutting incidents.

Simpson's literature review of self-mutilation, published in a collection about suicide, reports a 'very clear composite picture of the typical cutter' as being a young, attractive, intelligent woman. He mentions Offer's and Barglow's analysis, which ranges from the interpersonal and social settings to internal motivations. According to Simpson, they propose 'several motives such as gaining attention, the need to be loved and cared for, attempts to control aggression, tension reduction, and gaining prestige among the social group in the ward'. Such an explicit mention of the social setting merits significant disagreement: Simpson claims that '[e]lements of such motivations may well play a part in the dynamics of self-mutilation, but they are inadequate explanations – Why choose to gain attention or express the need for love by cutting one's wrist?'[92] Note that it is the social, communicative motivations singled out for their inadequacy, rather than (for example) the observations about aggression or tension.

However, he also mentions that many authors have focused upon issues of loss and abandonment as precipitants for cutting incidents. The social focus of this behaviour comes through most clearly in a passage where in Simpson discusses how 'cutting behavior can be learned and propagated in a hospital, clinic or institution' and how patients may compete for the title of 'chief cutter' through the number of stitches that they have received. In a startling description of the social significance of these acts, he writes: 'While patients may claim afterwards that they do not want others to know of their act, they often manage to flaunt the wound or their bandage like a newly engaged girl wearing her diamond ring for the first time'. There is clear ambivalence here, but he returns again to internal psychopathology in a passage referenced as central a decade later by influential scholar of self-mutilation, Armando Favazza. Simpson argues that 'self-mutilators commit what amounts to anti-suicide, employing the wrist-cutting as a means of gaining reintegration, repersonalisation, and an emphatic return to reality and life from the state of dead unreality'. This reintegration (drawing on the work of Karl Menninger) is conceived of as far more psychic than social. He claims that 'there exists a clearly identifiable condition of self-mutilation, usually involving wrist-cutting, which exhibits much of the stability of a syndrome'. He also asserts that '[w]hile self-mutilators represent a signif-icant problem group within the territory of suicide and para-suicide, they can be clearly distinguished from other similar presentations with significantly higher lethality, and thus warrant different treatment'.[93] The comparison with parasuicide shows how cutting and poisoning are increasingly seen as different phenomena.

Alec Roy's 1978 study from the Maudsley compares 20 consecutively admitted self-mutilating inpatients with a control group and explicitly attempts to rectify the failure of Gardner and Gardner (1975) to estab-lish difference between cutters and controls. He finds that nine self-mutilators reported anger at themselves as their predominant reason for cutting, whilst seven cited the relief of tension. He is unsure about the tension argument (even though Gardner and Gardner cite it as central) because 'the non-current cutter groups [those who had not cut within the 14 days preceding the interview] had anxiety and depressive symp-toms [, so] other variables may be important'. In formulating a general statement about self-mutilation, Roy considers 'intrapsychic, personality, interpersonal and psychosocial factors'. He expands on this, hypoth-esising that '[t]heir hostility, introversion and neuroticism may lead to anger and depression at their difficulties in forming and maintaining relationships and to the initiation and maintenance of this behaviour'. This roots the behaviour in the intrapsychic and personality realms, and makes the interpersonal and psychosocial distinctly secondary.[94]

Today, Keith Hawton is perhaps the best-known psychiatrist working on attempted suicide and self-harm in Britain. He is instrumental in establishing the Oxford Monitoring System for Attempted Suicide in 1976, and has written numerous papers on deliberate self-harm in a variety of ways, in a variety of settings – including studies at A&E departments and retrospective questionnaires in schools.[95] Research that he publishes in 1978, 'Deliberate self-poisoning and self-injury in the psychiatric hospital', is first mentioned in print in 1975, when clinicians from the Grayling well Hospital in Chichester refer to a paper given by Hawton in 1974. Hawton's figures are reported as referring to 'attempted suicide' in inpatients and day patients.[96] Hawton writes to the *British Medical Journal* in order to clear up possible confusion arising from such a citation. He states of his study of psychiatric inpatients: 'I would in any case be loath to use the term "attempted suicide" to describe the majority of these acts since many involved, for example, minimal cutting of the skin'.[97] Hawton is well aware that attempted suicide does not mean a genuine attempt at death in this context, but he is still strongly against calling 'minimal cutting of the skin' by the same name as a communicative act of self-harm that seeks help from the environment through the symbolism of suicide. For him, there is something different occurring, and it is no coincidence that this study is based in an inpatient institution – the Warneford Hospital in Oxford.

Hawton's study forms a bridge between the profile and description of self-cutting that emerges from psychiatric inpatient facilities, and the studies that include self-cutting as a minority behaviour in self-poisoning-dominated studies. It is published in 1978 and, throughout, it compares the inpatient data with the literature focusing on A&E studies (predominantly concerned with self-poisoners). The inpatient behaviours (named 'self-injury') are then differentiated from self-poisoning in terms of psychological motivation that are familiar from a twenty-first-century standpoint: 'the motivational factors leading to self-injury may be different from those underlying self-poisoning behaviour in the community...self-cutting is often used as a method of tension reduction and may be associated with states of altered awareness. Although self-poisoning may have a similar effect by temporarily interrupting consciousness, clearly the act is qualitatively very different'.[98]

What remains implicit in Simpson's choice of a control group of self-poisoners is made explicit by Hawton, who is able to expand upon the qualitative differences between self-cutting and self-poisoning – differences that are still included together without much comment in A&E-based studies. Hawton goes on to state that one patient reported that the difference is between feelings of tension (cutting) and feelings of hopelessness (overdosing): '[S]he cut herself in response to feeling extremely tense, and took an overdose when she felt depressed and hopeless'.[99] This maps quite precisely onto one of Jan Sutton's 'respondents' who is quoted as making the following differentiation nearly 30 years later:

> There was always a clear distinction for me between the cutting and the overdosing. The cutting was far more frequent and was about survival, about coping with the intolerable feelings I was carrying inside. Overdosing meanwhile was about giving up for good.[100]

Internal feelings as opposed to giving up for good – feelings of tension contrasted with feelings of hopelessness. It is a differentiation that has endured. Hawton (and Simpson, to a lesser extent) bridges the gap between inpatient studies of cutting and the presentation of self-damaging patients at A&E, giving the self-cutting minority of cutters in the general hospital samples the potential to be psychologically different.

But this again should not simply be viewed as a smooth progression with all researchers singing from the same song sheet. In 1979, the second edition of *Uncommon Psychiatric Syndromes* is published by David Enoch and W.H. Trethowan. Buried in the entry on Munchausen syndrome and related disorders (which involves the chronic fabrication or induction of illness in order to receive medical attention) is an intriguing passage and case study of one 21-year-old female. People who self-mutilate are said to 'scarify themselves with pieces of glass or metal, or indulge in parasuicidal wrist-cutting attempts'. This is said to indicate similarities with Munchausen patients in its 'tendency towards self-inflicted disability, together with a marked degree of attention-seeking behaviour and, perhaps, an unusual tolerance of pain and discomfort'. When the patient is questioned, she says, 'I sometimes feel I have to let the poison out that is in me!', and she also 'admitted to being very angry with herself and to feeling as if she were sitting on a volcano'. These are heavily internally focused motivations, and Enoch and Trethowan do mention that her behaviour 'undoubtedly reveals much of her basic emotional difficulties'. However, they also argue that 'such self-destructive behaviour ... must be seen, if it is to be understood at all, as a method of communication – a cry for help as well as for attention'.[101] There is nothing inevitably internally focussed about self-cutting behaviour, and it can be interpreted either as inward- or outward-looking. Once again, the meaning of behaviour is highly contingent.

Hugh Gethin Morgan's book *Death Wishes?* published in 1980, is based upon extensive study in Bristol. It shows how the assessments of hospital presentations of self-harm are changing. It is already noted how his studies during the early–mid 1970s use the term 'deliberate self-harm' to describe all methods of self-harm. Despite this terminological discussion, as we see at the beginning of this chapter, lacerations are not thought to be meaningfully different behaviour. All this changes by 1979. Here, Morgan argues for strong differentiation between self-poisoning and self-cutting, even though both actions are considered to have only an ambiguous relation towards self-accomplished death. He claims:

> In looking for causes of DSH [deliberate self-harm] it is important to consider self-laceration separately in order to discern psychopathological mechanisms which may be peculiar to it and which are not shared by those who take drug overdosage.[102]

Again, the importance of this claim can be seen retrospectively from the point in the early twenty-first century, where self-cutting and self-poisoning are significantly different. Morgan mentions a number of 'American writers' who tend towards a stereotype of a self-cutter being 'an attractive young woman' and suggest that self-cutting is 'in the nature of a schizophrenic psychotic reaction'. Like Myre Sim (see Introduction), he is unconvinced about the femininity of the stereotype, noting that '[o]ur Bristol survey demonstrated that, at lease in one provincial English city, men outnumber women amongst patients presenting at Hospital Accident and Emergency Departments following self-laceration' and thus the 'beautiful and female' stereotype is simply 'one amongst many'.[103] Thus there is both influence and distance from the American studies from British-based clinicians. Self-cutting emerges from its inpatient context and takes on renewed significance as a psychological object in its own right, whether presenting in an inpatient institution or at A&E.

Morgan sees self-laceration as concerned with an altered state of consciousness, a need to obtain relief from tension and a high incidence of obsessional, phobic and narcissistic tendencies. Immediately after this discussion, as if to restore a sense of balance, he states that 'DSH cannot be understood entirely in terms of intrapsychic pathology. There is a massive body of evidence testifying to its close relationship with interpersonal social events, and not merely as a blind reaction to them'.[104] Remembering that, for Morgan, DSH refers to both self-cutting and self-poisoning, it is clear that the behaviours are still linked, even if self-cutting requires a level of differentiation and discrete concern.

In 1982, the first edition of Keith Hawton's and José Catalán's *Attempted Suicide* is published. It contains a special chapter entitled 'Self-injury',[105] which details a tripartite division between: 'superficial self-cutting', which is 'usually of the wrist or forearm, associated with little or no suicidal intent'; this is followed by serious self-injury, which involves deep cuts that endanger blood vessels or tendons, as well as shooting, hanging and jumping from buildings which, as they note, 'are usually associated with serious suicidal intent'. Finally, there is the category of self-mutilation that 'may result in disfigurement' and is associated with psychosis; it may or may not be life-threatening. It is obvious that much nosological effort has been expended here. Differences are minutely examined and categorised in multiple ways. The authors mention that in Oxford, 'particular care has been taken to try to identify all cases of self-injury coming to the general hospital, irrespective of whether they have been referred to the hospital psychiatric service'.[106]

Hawton and Catalán note that wrist cutting has been treated as a distinct syndrome (referencing the North American literature), but they are unconvinced, adding that it is 'doubtful whether this is a useful approach to the problem, especially for clinical purposes'. They rehearse the now-familiar picture that '[t]he predominant sensation is one of tension, which steadily mounts until it becomes unbearable…Immediately before cutting, a sense of numbness or emptiness may be described'. Crucially, this differentiates the behaviour from self-poisoning. They argue that 'clinical teams which manage attempted suicide patients should be familiar with the special problem of patients who deliberately injure themselves, and not just deal with them as if the behaviour was the same as self-poisoning'.[107] This is exceptionally clear. In the second edition of their guide, they further note: 'Wrist-cutting, which is predominantly a behaviour of younger patients, is often repeated and in many cases appears to be a different phenomenon in psychopathological terms from self-poisoning'.[108] We are– so to speak – arrived at the present. And this context is that of the 1980s, where the relationship between the state and social life is being radically reimagined (rolled back), and where neo-liberal ideas of self-reliance and independence are dominant (see Conclusion).

The difference between inpatient and outpatient objects of self-harm

Having demonstrated that self-cutting emerges in certain (inpatient) places in British psychiatry and then is able to migrate and to transform analyses in other (general hospital) arenas, the final task is to ask why the inpatient and A&E objects are so different, despite people lacerating themselves presenting at A&E and many self-cutters also having taken overdoses. A significant part of the answer is to be found in the ways in which different therapeutic environments bring different behaviours to prominence. As noted above, self-cutting behaviour rarely becomes the object of intensive psychiatric scrutiny outside of inpatient institutions. Cutting only becomes scrutinised when inside the high-surveillance environment of a psychiatric inpatient ward.

When describing some implications of their Bristol study in 1975, Morgan et al. found that those patients who fell into the 'not interviewed' category were 'more likely to have lacerated themselves'.[109] In 1977 Norman Kreitman observes that 'there is little doubt that self-injury is under-represented' in the Edinburgh statistics.[110] Richard Turner and Hugh Gethin Morgan note in 1979 that casualty-department-based samples cannot be regarded as representative of all self-harmers, because it has been shown that '20% of all those who present to Accident and Emergency Departments were discharged home without being admitted to hospital, and these [so discharged] were younger and more likely to have lacerated themselves than those admitted to medical wards'.[111] The method of self-harm has practical consequences, as one escapes psychiatric scrutiny with greater regularity at a general hospital if cutting rather than poisoning.

Similarly, Hawton notes of his 1978 inpatient study that 'patients with minor scratches and cuts reported in this study might not have been referred to the general hospital and thereby identified if they had done this in the community'.[112] At A&E, however, Morgan reports that self-laceration might 'appear trivial when seen in hospital Accident and Emergency Departments'.[113] Conversely, self-poisoning figures regularly in the symptomatology of 'self-cutters' but is rarely emphasised. Waldenberg notes that many of his group of 'wrist-cutters' also took overdoses without intending to die.[114] Gardner and Gardner relate: 'We also had the impression that individual cutters took overdoses of drugs more often than the controls, but the actual number of cutters who had taken one or more overdoses was not significantly different from the control group'.[115] It is clear that the kinds of behaviour that come under psychiatric scrutiny in psychiatric hospitals and in community studies are very different.

The information that constitutes the inpatient studies relies heavily upon the levels of psychological and biographical scrutiny that only the inpatient setting can provide. McEvedy's dissertation provides a good example of this. The appendix of case histories he provides for each of his 'cutters' reveals the initial reason for admission of these patients. These reasons for admission include: 'hysterical paresis of the right leg', a suicidal attempt involving barbiturates, 'abnormal eating habits and loss of weight', 'difficulty getting along with people and depression', a referral from an Approved School because of depression and an incident involving severely 'slashed her wrists [and] an overdose of aspirin'. Another patient is referred to the Maudsley's forensic unit on a charge of shoplifting and was considered to be in need of admission because of an 'apparently sincere attempt at suicide by taking an overdose of tablets'. Another patient is admitted due to supposed temper tantrums and spiteful behaviour towards other children. The only patient admitted for self-cutting is Kay R., who came to be at St Francis because of the severely slashed wrists and an aspirin overdose – but the mention of the severity seems to preclude the kind of cutting in which McEvedy is interested.

However, once these people are inpatients, other behaviours are discovered retrospectively: Penelope E. is described as pulling the emergency cord on a train then presenting herself to train staff with cuts and scratches, which the police think self-inflicted. This case is apparently not referred for any kind of psychological attention. Similarly, after Kay R.'s admission doctors learn of a past surgical procedure to remove a needle fragment from her leg – allegedly the result of a fall over her sewing basket – an explanation the staff considers 'extremely unlikely' in retrospect. Whilst in the hospital she puts scissors in her mouth in a way that 'alarmed the nurses' and is referred to the psychiatrist, but does not end up seeing one.[116] This all points to how rarely certain self-damaging behaviours come under psychiatric scrutiny if performed outside of psychiatric inpatient settings.

J.B. Watson's patient is admitted for cutting – carried out in a particularly visible way. As Watson reports, the patient 'became anxious and depressed and began to cut herself with glass or razor blades on her arms and face. She was admitted to a psychiatric unit'.[117] It seems fair to assume that cuts on the face are much more noticeable (and perhaps more alarming) than those easily concealed on arms or legs. It is evident that very few self-cutters of the current literature cut themselves on the face, and facial self-mutilation is considered to be rather different to the kind of self-harm discussed here.[118] In 1972 Waldenberg is specific that only one of his patients is admitted due to self-cutting behaviour.[119] Again, this shows how behaviours such as cutting are much more likely to come to light once a patient is inside an inpatient institution, with all the opportunities for scrutiny (and perhaps desire for resistance of prescribed routines) that it entails. This could explain why this behaviour does not at first figure so prominently outside of these institutions. The environment is key in bringing to light certain forms of behaviour.

This may be (in part) because doctors do not believe cutting one's wrists to be a particularly dangerous act, in the sense that it rarely endangers life. McEvedy states that it is 'unlikely that death will result from a slash of the wrists' even though popular opinion 'continues to hold the belief that an injury [to the wrists] will prove rapidly fatal'.[120] In 1975, Simpson transforms this clinical view into numbers, as we have seen: '[A]ll acts of wrist-cutting were estimated at a lethality of 4 [lowest lethality]. There was a wider scatter of lethality scores for the self-poisoners, and an average score of 3.4'.[121] Incidentally, this is something also noted by the North American studies of self-cutting. One influential study observes that wrist cutting 'is an unusually difficult way to draw large amounts of blood', whilst another claims that 'wrist slashing' is 'a notoriously poor method of suicide'.[122] This means that people presenting at hospital are less likely to be admitted, and that cuts on the arms (but not the wrists) are unlikely to be discovered, let alone be the cause of a trip to A&E. However, once a clinical object is established in inpatient facilities, it can travel and become a psychologically distinct category, into which A&E patients might fall.

Concluding thoughts

Self-cutting emerges as an epidemic phenomenon and a management problem in psychiatric inpatient institutions, and it shifts from these sociologically informed perspectives towards an approach more focused upon an internal psychopathology which involves intolerable psychic tension. After this has become stable, it migrates to A&E departments, and informs analyses of the small numbers of self-cutters who present there. Thus, self-cutting and self-poisoning, treated as largely similarly motivated in A&E studies in the early 1970s, are strongly differentiated by the end of that decade.

This chapter charts the changes in explanations for an emergent mental health problem in Britain during the 1960s and 1970s. The behaviour, which is called self-cutting, wrist-cutting, self-harm and self-mutilation, does feature in the studies that focus upon self-poisoning, but is largely ignored as a methodological quirk, as is shown at the outset of this chapter. This contrasts with the emergence of cutting in psychiatric inpatient institutions – which first figures as a management and behavioural epidemic problem. The management problems (and the negative reactions of some staff members) feed into an emphasis of internal, psychopathological aspects of the phenomenon, rather than the communicative, imitative and competitive (potentially 'manipulative') aspects. When this has sufficiently stabilised, it is able to inform studies based at general hospitals. Explanations emphasising internal over possible external factors find traction when compared to the self-poisoning at A&E departments.

Self-cutting also receives a boost in visibility when it features in the American Psychiatric Association's Diagnostic and Statistical Manual of Mental Disorders for the first time in 1980 (DSM-III), as a possible symptom of borderline personality disorder. Hawton and Catalán mention 'personality disorders' in their analysis of self-mutilation in 1982 and, in the second edition of the same text (1987), they add the following sentence: 'In the USA the DSM III diagnosis of "borderline personality disorder" would often be used for such individuals'.[123] This implies that the DSM and borderline are involved in the increasing prominence of such symptoms – as self-mutilation features in the DSM for the first time as a symptom of borderline in 1980. By 2014, it is afforded a diagnosis of its own: Non-Suicidal Self-Injury.

The behaviours of self-poisoning and self-cutting emerge in very different institutional settings, despite their common co-occurrence in the same patients. Self-poisoning is much more likely to bring a person under medical scrutiny, whereas superficial cuts to the arms are much more likely to be noticed when a person is already in an inpatient setting. This gap between inpatient and A&E studies has been forgotten in the transformation in the visibility of the two behaviours of self-cutting and self-poisoning.

Except where otherwise noted, this work is licensed under a Creative Commons Attribution 3.0 Unported License. To view a copy of this license, visit http://creativecommons.org/licenses/by/3.0/

OPEN

Conclusion: The Politics of Self-Harm: Social Setting and Self-Regulation

Almost three decades ago, historian Howard Kushner writes of his unease at increasingly neurological understandings of behaviour such as suicide. He argues that '[o]ne feature of neuropathological approaches, however, seems unaffected by this increasing sophistication: the more scientifically complex these investigations become, the more they tend to ignore the social and historical context in which the behavior that they seek to explain takes place'.[1] In these accounts, neurology displaces social context. In characteristically forthright terms, in 2014 Roger Cooter describes the turn to neurological explanations as 'like becoming the victim of mind parasites' because these explanations foreclose the ability to think critically about the social and cultural context of the explanations themselves: they are presented as universally true and outside of culture or history.[2]

Self-harming emerges as an epidemic in Britain as pathological social communication and is transformed into affective self-regulation. Therefore it can serve as a barometer of broader changes in understandings of human behaviour. Self-harm in the form of self-poisoning is understood as highly social; the self-damage by self-cutting, which displaces it, is understood as predominately internal. It is this internal, emotional quality that enables its easy fit within neurological and neurochemical frames of reference. The career of these behavioural archetypes can tell us much about the dominant ways in which human behaviour is understood: the ways in which behaviour is given meaning according to cultural assumptions that shift in their relative influence and credibility. In this conclusion I seek to do five things. First, I recap in summary form, the book's main arguments and content. Then I want to reflect a little on the book's methodological underpinnings, to show that I do not exempt myself from the kinds of analysis carried out in the book. This happens in the Conclusion rather than the Introduction because it is easier to reflect in a comprehensible way upon this process when the argument has been laid out. Third, I sketch (very briefly) some of the ways in which self-harm-as-affective-regulation is now within the orbit of neurological explanations. Fourth, I expand upon the political significance of the internal, emotional understandings of self-damaging behaviour. Finally, I reflect upon the implications of this historical account of self-cutting and self-poisoning for human behaviour in general.

Summary of argument

Self-poisoning as pathological communication has a relatively short shelf-life. In 1975, Eliot Slater produces an article describing the state of psychiatry in the 1930s. He observes that '[t]he young in those days did not have today's facilities for drug addiction, for self-inflicted wounds, for attempted suicide as a "cry for help"'.[3] What seemingly starts as a comment on the increased level of drugs circulating in 1970s society strikes a much more profound note by the end. In the 1930s, the 1970s patterns of 'attempted suicide as a cry for help' are simply not available. In the twenty-first century, whilst not invisible, self-poisoning as a cry for help has been eclipsed by deliberate self-harm, based around self-cutting for emotional self-regulation.[4]

As we have seen, between the 1930s and 1970s a number of objects under a variety of names (attempted suicide, pseudocide, self-poisoning, parasuicide) emerge through traffic between the therapeutic approaches of general and psychological medicine. Throughout the middle third of the twentieth century the relationship between psychological and general medicine is reconfigured, and the concepts used to label, treat and analyse patients presenting at hospital with a self-inflicted physical injury are subject to much change. Actions configured around violence and a fear of imminent fatal repetition give way, slowly and unevenly, to actions interpreted as a result of childhood psychological trauma, or attempts to communicate social and domestic stresses. This is not just a change in interpretive strategy, with some form of object constant beneath these different responses: the objects are fundamentally reconstituted in different contexts, by different practices.

The police-watching controversies articulate a concern over 'would-be suicides' due to a financial dispute between the police and voluntary hospitals. The potential for violence and repetition is emphasised as part of a strategy by hospitals to compel police to remain in attendance whilst the patient is treated. The potential for immediate repetition carries with it the implication that the attempt is aimed at death. A dispute then emerges between workhouse infirmaries and voluntary hospitals that again emphasises violence, but this time in order to place 'attempted suicide' within the remit of workhouse infirmaries, as they are supposedly better equipped to deal with mental patients.

Legislative changes in 1929 and 1930 abolish the Poor Law and promote the informal (non-certified) treatment of mental disorder. As a result, psychological and general medicine come into a closer relationship around (mental) observation wards attached to general hospitals. In many cases these wards are the old workhouse infirmary mental blocks, with workhouse infirmaries turned into local authority hospitals at the abolition of the Poor Law. This closer relationship gives Consulting Psychiatrist Frederick Hopkins consistent access to various 'physically injured' patients brought to his Liverpool observation ward. This arrangement makes visible a broadly coherent group of people whom he deems to have attempted suicide due to various social and constitutional factors, including 'domestic stress'. He is aware of, but equivocal about, an old notion that attempted suicide is principally a manipulative communication.

The engagement with the psychological casualties of the Second World War prompts a number of interpersonal therapeutic experiments. Psychological problems and mental suffering are seen as inseparable from factors in the social environment. As part of this process, therapeutic communities are established at various sites in the United Kingdom by psychiatrists including Tom Main, Maxwell Jones, Wilfred Bion and John Rickman. Social environment and psychopathology become ever more closely entangled.

In 1948 the NHS is inaugurated, with mental health included in the comprehensive service. This removes any disputes about payment for certain classes of patient and effects a closer connection between general and psychological medicine. It is also part of increased collective and social welfare provision, nationalised industry and centralised planning. The remit of the state to manage, fund and direct social life (through social work, child protection, child guidance, welfare requirements and so on) is expanded. As part of this shift towards collective provision, the connection between mental and physical medicine is strengthened. At accident and emergency (A&E) departments for cases of attempted suicide, this link is not sufficient to produce a social constellation around a physical injury conveyed to hospital. The presence of psychological medicine is still too marginal in casualty departments, where the overwhelming focus is acute somatic medicine. However, in the early 1950s facilities for the treatment of poisoning, psychological scrutiny and psychiatric social work (PSW) expertise all converge at an observation ward in Edinburgh. This results in psychological scrutiny of physically injured patients, but also in the rooting of psychopathology (through the conceptual apparatus of John Bowlby) in childhood emotional deprivation in so-called broken homes. Psychiatrist Ivor Batchelor and PSW Margaret Napier operate in tandem to construct a vision of psychological maladjustment and low stress tolerance in the background of these attempted suicide patients. This is largely achieved through intensive questioning and assiduous follow-up by PSWs. A similar object of concern is publicised around the same time in London observation wards by Erwin Stengel and co-workers (principally PSW Nancy Cook). This attempted suicide is again part of a crossover between mental and general medicine, but more focused upon a present-centred (often unconscious) appeal, in response to social difficulties.

In the late 1950s the final legal impediments to psychological treatment at general hospitals are swept away in the Mental Health Act (1959) as part of a wider effort to eliminate as far as possible the differences between the treatment approaches. Connected to this effort, and using Stengel's research, suicide and attempted suicide are decriminalised in England and Wales in the Suicide Act (1961). Both of these acts remove legal machinery from areas considered psychological in nature. Thus, to ensure that the appropriate kind of care is forthcoming, they are swiftly followed by government guidance to hospitals recommending psychological assessment for all attempted suicide cases seen at accident and emergency departments. This is actively followed up by the Ministry of Health; the variable results recorded demonstrate the difficulty of focusing intensive psychological scrutiny at casualty departments.

Whilst the government passes legislation, the Medical Research Council (MRC) sets up a unit for psychiatric epidemiology that ends up in Edinburgh, at the same ward that produces some of the 1950s studies. With the MRC's backing, Neil Kessel embarks upon a project to study 'attempted suicide', which he renames self-poisoning. Collaborating extensively with PSWs, Kessel roots the causes of self-poisoning firmly in the present, and as a conscious appeal, in an all-encompassing category of distress, centred upon a feminised vision of the home and supposed marital disharmony.

As the government starts to run down the asylum system and promote psychiatric units in general hospitals, a large number of studies, with varying degrees of psychological scrutiny, are able to effect the transformation from 'physical injury' at casualty to 'socially rooted appeal'. The growing self-evidence of the social constellation (in a society where the state's social responsibilities are much larger than today) remains a product of much intellectual and practical effort. It means that a broadly causative social setting is increasingly presumed around a casualty admission for poisoning. This presumption makes the behavioural category increasingly stable, public and available as an intelligible human response to interpersonal difficulties. This broader self-evidence fuels new terminological offerings, with 'parasuicide' the latest neologism, proposed in 1969.

Alongside (and entangled with) this story runs that of self-cutting from the early 1960s. Self-cutting (especially of the wrists and arms) has long featured as a seeming methodological quirk in self-poisoning and parasuicide studies, presenting at hospitals as approximately 5% of cases of self-damage. In these general hospital-based studies, self-cutting or self-lacerating are not seen as motivated differently to self-poisoning. However, self-cutting also emerges in the context of psychiatric inpatient institutions. Influenced by North American psychoanalytic inpatient literature, a British corpus of studies on self-cutting, self-mutilation or wrist-cutting emerges. This is initially seen as related to the strictures and constraints of the inpatient environment and provokes much interest and concern due to its highly distressing and contagious epidemic nature. However, as the 1960s pass into the 1970s, a sense emerges from these inpatient studies that self-cutting is motivated by internal, emotional psychopathology grounded in a sense of intolerable psychic tension.[5] This sense remains strong in the current literature on self-cutting. As Karen Skegg reports in 2005 in the *Lancet*, this is not a clear-cut disavowal of communication, but instead the relative dominance of internal, emotional and tension-based factors: 'Reported motivations for adult superficial self-mutilation included: to relieve tension, to provide distraction from painful feelings, as self-punishment, to decrease dissociative symptoms, to block upsetting memories, and to communicate distress to others'.[6]

This re-reading of self-cutting is then (from the late 1970s) imported from the inpatient studies into the A&E-based samples. This first seems to be done coherently and explicitly in a study conducted by Keith Hawton in the mid-1970s and published in 1978. This differentiation then begins to make sense of the hospital/A&E presentations during the 1980s. Thus the idea that self-cutting and self-poisoning are differently motivated behaviours begins to gain traction. Self-cutting becomes stabilised as a method of internal affective regulation, whilst self-poisoning is rendered more ambiguous: it features both as a genuine suicide attempt and socially directed self-damage.

The growing ambiguity of self-poisoning and its eclipse by self-cutting might tentatively be connected to a fracturing of the kind of psychological expertise that first produced it. Between 1977 and 1980 a number of clinical studies are published that question whether assessment by psychiatrists is necessary in every case. A consensus is reached that other professionals such as social workers, nurses or junior doctors – with some training provided by psychiatrists – are equally competent to do this.[7]

In 1981 a working party is set up by the Royal College of Psychiatrists at the invitation of the Department for Health and Social Security (DHSS, successor to the Ministry of Health). The group includes Norman Kreitman, Hugh Gethin Morgan and Irving Kreeger, who are asked to review government guidance on the management of deliberate self-harm. (This term is defined in the report as covering 'patients who injure themselves by poisoning or other means' – so poisoning is predominant here.) One of the outcomes of this working party is that a Hospital Notice is issued in 1984, drawing attention to the report included with it. Specifically, the notice emphasises the recommendation that 'suitable trained medical practitioners, other than psychiatrists, may undertake the psychosocial assessment of patients who deliberately harm themselves, and that referral – in some cases, to professional workers, other than psychiatrists, who have received special training – may be considered appropriate'.[8] Thus, right at the point where self-poisoning treated at hospitals is becoming differentiated from self-cutting, the assessment of the (largely) self-poisoned patients at A&E can be delegated away from psychiatrists.

This fractures the intense psychiatric scrutiny (based around research articles) that has been shown to be so important in stabilising socially directed self-poisoning as a self-evident object of enquiry. So here again are specific practices that are prescribed by the state, and that influence the visibility of a psychological object. As much as the rise of self-cutting might resonate – somewhat perversely – with neo-liberal ideas of self-reliance, and be part of the retreat of the state from social welfare spending (which is further explored below), there are still specific, mundane, administrative practices that correspond to the retreat of self-poisoning from national significance.

It is not coincidental that the texts at the forefront of raising awareness about self-cutting do not normally come from A&E, but instead from psychological clinicians involved in counselling – where the potential for intensively scrutinised case studies is much higher than in busy hospital casualty departments. In addition, visibility is granted to self-cutting through its inclusion as a symptom of borderline personality disorder in the third edition of the *Diagnostic and Statistical Manual of Mental Disorders* in 1980.

Methodological reflection

This story is told through two principal sources of information: articles in medical and psychiatric journals, and government documents at the National Archives, Kew, and at the Lothian Health Services Archive in Edinburgh. In the first category there are also some dissertations – principally McEvedy (1963), Keller (1970) and Waldenberg (1972) – that are not in journals, but still adopt the style, tone and formality of psychiatric research. The sources at the National Archives provide the basis for discussions of the police-watching controversies of the early twentieth century, the machinations around the Suicide Act and Hospital Memorandum (1959–61), some extra information on Kessel's Edinburgh unit (1961–5), and the hospital notice of 1984. The Lothian Health Services Archive principally furnishes extra information about Ward 3 of the Royal Infirmary of Edinburgh.

Although archival sources form a significant proportion of this book's basis, the predominant source base remains scholarly psychiatric research articles. Because these are written with scholarly apparatus (chiefly references), I am able to follow acknowledged trails of thought and influence. I use these to construct a more or less established 'canon' of documents by authors who refer to each others' work. The key names should be familiar – Hopkins, Batchelor and Napier, Stengel and Cook, Kessel, Kreitman. This means that much of the rise (and fall) of self-poisoning is seen through the lens (and constraints) of research output. When this significantly drops off, and self-cutting largely displaces it as the meaning of 'self-harm' or 'self-damage', this does not mean that I am making a strong argument about the numbers of people performing these actions. As I hope is clear, the numbers of people reported in any given study depend largely on the institutional basis for such a study (for example, whether inpatient or outpatient) and on the specific practices used to find and evaluate cases.

This historical method is paired with very tight focus on the subject matter of the articles: unpicking lines of argument, searching for mentions of specific practical arrangements, evaluating the position of various professionals (PSWs, police, etc.). It does not leave very much space for the 'patient experiences' of self-cutting or overdosing. This features in a small way, when patient testimony is used and deployed by psychiatrists as evidence. This is especially useful when patients confound expectations (as in Watson's 1970 study), or requires significant intellectual work to make it fit (as in Waldenberg, 1972). However, this is principally a study of specific hospital practices, a certain set of psychiatric ideas about the social setting, and how these might resonate with a wider political context: a shift from a welfare-based, socially interventionist consensus to one of individuated, market-oriented competition. Roy Porter champions ideas of the 'patient's voice' as central to the history of medicine, but this is not my principal area of interest.[9] I am far more concerned with how ideas and research practices interact and produce the concepts and shorthand that humans use to understand themselves and others. Basing this book on the experience of the patient would make it a very different project. In addition, Joan Scott writes persuasively:

> When experience is taken as the origin of knowledge, the vision of the individual subject (the person who had the experience or the historian who recounts it) becomes the bedrock of evidence on which explanation is built. Questions about the constructed nature of experience, about how subjects are constituted as different in the first place, about how one's vision is structured – about language (or discourse) and history – are left aside.[10]

I am most interested in how 'vision is structured', in how ideas and practices come to influence what is possible and explicable behaviour, and how these change. This is not to demean patients or their stories, experiences or identities, but to say that this history attempts something different. The patients and their experiences recede in this telling, as do the psychiatrists to an extent. What is left are practices, arrangements, ideas, concepts – all the things that recur in psychiatric journal articles and government documents. This, like all history, must resemble its sources, but remains useful – hopefully to people other than myself – because it enables new connections to be made around self-harm, society, psychology and politics. It might make the various individuals involved in the story less visible (in terms of their experiences), or flatten them out to their research contributions, but it also allows new links: between categories of identity and the rise of professional groups; between broad political contexts and clinical categories; between an intellectual climate in psychology and psychiatry and the ways in which we understand self-damaging behaviour; between politics and the ways in which people understand themselves and their identities.

The rise of neurology

To return to specifics, we see that self-cutting is both a residual and a newly emergent category. It is understood – gradually and unevenly – as a method of affective self-regulation rather than social communication. This opens the way to neurological explanations of the behaviour. This happens because neurological explanations focus upon the individual's nervous system as a privileged site of understanding. The reason that it is only a small step from 'individual tension' to 'neurochemistry' is that both approaches, or concepts, take the individual at their starting points. A communicative attempt, in contrast, focuses upon a social situation in which various people are embedded. However, even this contrast has recently become unstable, as there is work that investigates the 'neurology of social cognition' as well as sociological work on the discipline of neuroscience (upon which this book has drawn).[11] However, the point stands that internal emotional turmoil maps much more easily onto neurological understandings than does psychosocial communication.

Although the neurochemistry of complex behaviour is widely acknowledged to be in its infancy (a claim also made in the mid-1970s), there are a number of guiding principles that underwrite these perspectives. Regardless of the particular system or neurotransmitter that is implicated, these sorts of studies are all based around the assumption that neurochemistry is at the root of the behaviour, and operates prior to culture, and is indeed, outside culture. As Hilary and Stephen Rose argue with respect to molecular biology: 'Again and again the molecular biologists leading the sequencing [of] the human genome [between 1990 and 2013] claimed that the completed genome would constitute human identity'. They add that 'The neurosciences have not been left behind; their claims to explain selfhood, love and consciousness as located in certain brain regions...have been articulated in a string of popular books'.[12] Given the audacity and ambition of these claims, it is unsurprising that neuroscience and neurobiology are increasingly utilised to investigate the (comparatively modest-sounding) self-cutting-as-tension-release in order to reveal its neurological basis.

Michael Simpson is among the first to speculate upon a biological basis for the behaviour of self-cutting in 1976, but he is notably cautious in ascribing the behaviour any secure biological basis.[13] In 2001, Fiona Gardner (a psychoanalytically trained therapist) writes in a cautious and equivocal vein about 'self-harm':

[T]he behaviour can be coercive, in that self-harming produces a wanted response from others; second, it is relieving, in that the action produces a lightening of mood, either through biochemical alterations and the associated release of endorphins (the body's own analgesics), or conditioning, or symbolically.[14]

She mentions social ('coercive') explanations, followed by biochemical, classical conditioning and symbolic understandings. These cautious studies exist alongside publications by such as Schroeder, Oster-Granite and Thompson's text, *Self-Injurious Behavior: Gene-Brain-Behavior-Relationships*, published by the American Psychological Association in 2002, and which pushes much harder to understand behaviour in terms of the brain and genetics.[15] In the 2010s, there is increasing recourse to neurochemical and biological explanations to explain 'self-injury'. David Brent, in a recent editorial for the *British Medical Journal*, sees self-injury as 'most commonly used as a mood regulation strategy...thought to relieve negative affect through the release of endogenous opioids'. Whilst Brent does argue for the relevance of the social context, he also maintains that the difference between suicide and self-injury can be established with reference to neurochemicals: 'Although nonsuicidal self-injury and suicide attempts often occur in the same individual and share some common risk factors, their motivations, reinforcers, and neurobiology are distinct'.[16] This is an explicit attempt to separate suicidal behaviour and self-injury, not simply in terms of motive but in terms of a distinct neurobiological pattern. In addition, as emotions and moods are increasingly understood in neurological terms, so self-injury becomes enmeshed in neurological explanations. For example, as borderline personality disorder (closely associated with self-injury) is understood through neurochemical frames of reference (e.g., neuropeptides), self-injury is increasingly 'neurologised' by association.[17] Efforts have also been made to associate self-injury and suicidal behaviour with the neurochemical serotonin and the serotonergic system.[18]

However, there also exists considerable circumspection amongst the more prominent psychological experts on self-injury about how much this behaviour might be reduced to biological bases. Favazza is sceptical of neurology and neurochemistry – as might be expected of a 'cultural psychiatrist' whose undergraduate degree is in anthropology under Margaret Mead. He notes that there may be a swing back towards analyses that focus upon the social or cultural environment: 'although psychiatry is focused on the primacy of cellular, genetic and neuronal approaches, there is a growing recognition that culture cannot be ignored'.[19] To him, it is clear (in 2011) that cells genes and neurons are at the forefront of conventional explanations. In a similar vein, sociologists Adler and Adler are clear about their desire to 'demedicalise' self-cutting, understanding it instead through sociological concepts such as deviance and social reinforcement.[20] It should be noted that sociological and psychological explanations persist – based upon learning and peer-group influence, and yet remain based upon ideas of emotional regulation. This is not a simple dichotomous split. However, over the past decade there have been many efforts to understand self-harm through neurochemical and neurological frames of reference. Health communications scholar Warren Bareiss concludes that media narratives of self-injury consistently downplay possible social causes of self-injury in favour of a model that understands self-injury as a personal choice.[21] This idea of an individualised, personal choice meshes well with neurochemical understandings, as well as with market-based ideology that is centred upon a rational, autonomous consumer.

This is a complex and nuanced picture, where social – and sociological – explanations can co-exist with ideas of internal tension and can also feed into neurological explanations. There is no easy way to sum them up. However, we can be more certain about the shifts at the heart of this book: that the archetypes of self-damage from the 1930s to the 1980s have undergone radical transformations. This corresponds to local, mundane and administrative innovations, but also feed off and feed into much broader political constellations. It is to these that we now turn.

Neo-liberalism, individualism and biomedicine

The broad political picture in the United Kingdom between 1945 and the end of the 1970s is conventionally thought of as characterised by consensus politics, commitment to welfare and significant nationalised (collective) ownership of industry (including transport, communications and health care). Such a collective outlook corresponds with the 'local picture' drawn in this book, consisting of a psychological perspective that is acutely aware of collective social life, communication and the embedded nature of human beings in their particular social contexts. This consensus politics is displaced from the late 1970s by a world view in which the family retains its importance, but there is much more of a focus on individual competition and self-discipline.

In general, neo-liberal thinking is based upon the primacy of market forces and the desirability of individual competition. Efforts to provide collectively are seen as stunting the individual's competitive edge. This perspective on human life has its roots in the political philosophy of Friedrich von Hayek and the economics of the 'Chicago school', linked with Milton Friedman and his associates and students. In Britain, this approach is most closely associated with the three governments headed by Margaret Thatcher between 1979 and 1990. It has been labelled 'neoliberalism' due to its stress on the old liberal values of economic freedom and self-reliance. Roger Cooter describes it as 'the anti-Marxist philosophy-cum-ideology founded on a view of human nature as entirely self-interested and incapable of thinking beyond "the market," which it constructs and sells as an autonomous force'.[22]

Journalist Andy McSmith's popular history of the 1980s, *No Such Thing as Society*, takes as its title the immortal words uttered by Margaret Thatcher in an interview with *Woman's Own* magazine in 1987. In full context, Thatcher argues:

> I think we've been through a period where too many people have been given to understand that if they have a problem, it's the government's job to cope with it. 'I have a problem, I'll get a grant'. 'I'm homeless, the government must house me'. They're casting their problem on society. And, you know, there is no such thing as society. There are individual men and women, and there are families. And no government can do anything except through people, and people must look to themselves first. It's our duty to look after ourselves and then, also to look after our neighbour. People have got the entitlements too much in mind, without the obligations.[23]

This is an exceptionally clear message that social problems (such as homelessness) should no longer be the preeminent concern of the state but of individuals (and in fact the individuals who are homeless). It displays a clear shift from a governmental responsibility for social problems to individual responsibility. We move from the social to the individual. This can be usefully contrasted with Erwin Stengel's concern, quoted at the start of Chapter 3, about a 'society which has made every individual's welfare its collective responsibility'. In one sense, Stengel and Thatcher are worried about the same thing: the burden that the exploitative 'few' might exert on the hardworking 'many'. However, Stengel seems broadly to accept such a state, whilst Thatcher seeks to dismantle it.

As McSmith makes plain, this is an economic policy as much as anything else. Thatcher is congratulated upon the Conservative victory in 1979 by economist Milton Friedman (soon to become policy advisor to Ronald Reagan) in a rather grandiose exchange. Friedman gushes: 'Britain can lead us all to a rebirth of freedom – as it led us all down the road to socialism'. Thatcher replies: 'The battle has now begun. We must win by implementing the things in which we believe'.[24] McSmith also makes abundantly clear that the 'no such thing as society' sentiment is present in Thatcher's thinking in the late 1970s (and not just the late 1980s when it appears). He cites handwritten notes for a 1979 speech proclaiming 'no such thing as collective conscience, collective kindness, collective gentleness, collective freedom'.[25] Her abhorrence of the collective and the social, and her championing of the individual, maps well onto the shift from social communication to internal emotional regulation.

Michel Foucault's lectures at the College de France in the late 1970s contain a sophisticated discussion of neo-liberalism and its significance. He notes that it emerges in two distinct places, in similar forms: a German form, linked to a critique of Nazism and the post-war reconstruction and an American form, defined in opposition to Franklin D. Roosevelt's 'New Deal' and the federal interventionism of the Democratic presidential administrations of Harry Truman, John F. Kennedy and Lyndon B. Johnson. Foucault argues that these forms are united by 'the main doctrinal adversary, [economist John Maynard] Keynes... [and] the same objects of repulsion, namely, the state-controlled economy, planning, and state interventionism'.[26] According to Foucault, the state is reconceptualised as the guarantor of economic freedom, but more than that:

> Since it turns out that the state is the bearer of intrinsic defects, and there is no proof that the market economy has these defects, let's ask the market economy itself to be the principle, not of the state's limitation, but of its internal regulation from start to finish of its existence and action. In other words, instead of accepting a free market defined by the state and kept as it were under state supervision... [instead] adopt the free market as organizing and regulating principle of the state, from the start of its existence up to the last form of its interventions... a state under the supervision of the market rather than a market supervised by the state.[27]

Given the huge range of commentators on this shift, I here take just three other examples almost at random, for illustrative purposes. Perry Anderson's collection, *Spectrum*, analyses the writings of diverse political thinkers, and his appraisal of Ferdinand Mount (active in writing

Conservative Party policy in the 1980s, including the General Election Manifesto of 1983) is that the Labour Party's 'construction of a welfare state, technocratic in design and bureaucratic in delivery...is the consistent object of Mount's dislike'.[28] Butler's and Drakeford's analysis of social-work scandals in twentieth-century Britain characterises the Conservative governments of the late 1980s and early 1990s as 'redefining the welfare state in such a way that a premium was attached to notions of individual rights and personal freedoms'.[29] Rose and Rose argue that successive governments from the late 1970s onwards begin 'enthusiastically exchanging the political economy of the welfare state for that of neo-liberalism. The rise of transnational corporations, able to spread production processes across countries...together with the attack on organised labour, began to sap the very foundations on which the welfare state was built'.[30] Here we have the core of neo-liberalism: individual rights, antipathy towards the welfare state and organised labour, and a stress upon self-reliance rather than collective provision.

This political shift broadly coincides and intimately corresponds to the much more individualistic reading of self-damage, based upon emotional self-regulation. Indeed, neo-liberalism's stress on individual actors' radical freedom to make choices for their own benefit fits well with a model of self-harm that emphasises the individualistic, private feelings of tension, and the self-regulation of these through cutting. The coincidence of neo-liberal political ascendency from the early 1980s in the United States and United Kingdom, and the displacement of the social setting from understandings of self-damage are not chance occurrences.

In a similar vein to this book, Gillian Harkins analyses the shift from a welfare economy to neo-liberal one in terms of the emergence of certain human categories of behaviour. She connects the socio-economic shift to neo-liberalism to an emergent concern about predatory paedophiles:

> Harkins links the way in which we have constructed the paedophile as the ultimate monster to the vilification of the state as the enemy of the free market. Each form of discourse is linked to, and helps support the other, through a shared model of human nature and its interests. The state and the paedophile [are] depicted as stealing the natural potentiality of the child.[31]

So here we can see the ways in which apparently independent phenomena are linked to broad political changes, perhaps counter-intuitively. The figure of the paedophile and the new vision of a suffocating state are founded upon the figure of the child and its potential. Harkins expands upon this link, claiming: 'Older modes of social security ... will be replaced by demands for a new type of "security" in the face of universal danger [the paedophile]. This security operates through the proliferation of risks and controls rather than the enclosure of disciplinary space'.[32] Thus Harkins sees a shift from enclosed, family, social spaces, to a much more diffuse and distributed space. This is analysed and developed through the varied categories and objects that populate our lives. In the same way, the rise of self-cutting as based upon autonomous, self-regulating individuals pushes out a reading of socially embedded, collective responsibility for psychological distress. No longer is pathology redistributed onto spouses or social relations (in Thatcher's terms 'casting their problems on society') – it is internal, individual and self-regulated.

The relationship between the two outlooks and their different scales (macro and micro) is complicated and rather opaque. It is approaching the banal to say simply that they feed off each other and correspond to each other. This sentiment might be developed by arguing that from the infinite possibilities of human behaviour, only a small number ever congeal into perceptible objects and are labelled as traits, syndromes or patterns. We see with self-cutting that a large number of other behaviours (such as swallowing objects, smashing windows, paroxysms of rage or social imitation) are consistently downplayed in order to produce a comprehensible object. In the same way, self-poisoning as communication neglects internal psychological states in favour of charting the psychological significance of the environment. In a general sense, these objects rely upon the intellectual and institutional conditions, where they are studied and from where they are publicised (secure inpatient facilities, A&E, counselling services, psychoanalytic interviews, and so on). The objects that appear from these settings can then be regulated, studied or managed (by government memoranda, informal referral arrangements, or specially designed questionnaires). If that management is removed or undercut (by a rethinking of the responsibilities of the state), then the objects fall from view, leaving space for others. These new objects are more likely to attain prominence if they resonate with other changes going on in the political sphere.

But it is important to remember that mundane arrangements like the fracturing of psychiatric scrutiny on self-poisoning in 1984 are just as important – and certainly more accessible to historians who seek to explain change. In a similar vein, Waldenberg's (1972) refocusing attention away from communication (and its association with 'attention-seeking') is part of his strategy for dealing with nursing staff's 'grumbling' about self-cutting patients. In order to promote the care of these patients in ways he considers appropriate, he emphasises internal tension. These arguments remain important today, in the politics of deliberate self-harm. Labels such as 'attention-seeking', bandied about by the media, tend to trivialise the behaviour, so clinicians who are interested in taking it seriously and treating it might become wary of discussing or emphasising 'communicative intent'.[33]

We have discussed the resonance that self-cutting as tension-release has with neo-liberalism, and also that between self-cutting as tension-release and neurochemistry/biomedicine. We can complete that particular triangle of associations by briefly discussing the resonance between technological biomedicine and neo-liberal economics. As Kaushik Sunder Rajan argues in *Biocapital*, 'the life sciences represent a new face, and a new phase of capitalism and, consequently biotechnology is a form of enterprise inextricable from contemporary capitalism'.[34] Rose and Rose follow a similar line of thought, arguing that 'the life sciences have been transformed into giant biotechsciences, blurring the boundaries between science and technology, universities, entrepreneurial biotech companies and the major pharmaceutical companies, or "Big Pharma"'.[35] So the welfare state is rolled back, individualism and self-reliance are stressed, capitalism becomes largely unregulated, biotechnology flourishes and, self-damage as response to a social setting is displaced by self-damage as self-regulation of internal tension. This internal tension is then significantly (though not totally) 'biologised' and rooted in brain biochemistry. A detailed study of the interactions between these threads is for another book. What I want to emphasise here is that the mass of labels and psychological objects that populate our lives do correspond to wider political contexts. Self-poisoning as a cry for help is largely invisible today (even as self-poisoning numerically dominates A&E statistics for self-harm) partially because the embedded, funded and self-evident awareness of social contexts has largely disappeared from the political mainstream.

Neurology and neo-liberalism are also linked through their wholesale forgetting or belittling of the social context. In fact, it is this determined omission that gives both neo-liberalism and the turn to neurology their fundamentalist zeal. It serves as fuel for its evangelising of the revealed neurological, or competition-based eternal truths of human nature. To quote Rajan again: '[C]apitalism, which is triumphantly acknowledged today as having "defeated" alternative economic formations such as socialism or communism…is therefore to be considered the "natural" political economic formation'.[36] This is not to belittle neuroscience or capitalism in a tit-for-tat battle as neurologists and neo-liberals attempt to cast 'the social' as irrelevant. Nor is it to claim that neurochemistry has no impact upon how humans behave: Who could doubt the influence of the body upon the mind? The thing that baffles and unnerves me in equal measure is the refusal of some to countenance that this embrace of neurology is itself a culturally, socially situated phenomenon. The ways in which we search for a handle on human nature change over time, and are parts of humanity's socially influenced, culturally saturated existence. The claims of science to be beyond culture, to be a method by which unarguable truth is revealed, begins to sound more and more theological the more entrenched it gets. It also fails to see how the ways in which science considers itself beyond cultural contexts and biases – the complex notion of objectivity – have themselves changed over the centuries.[37] The idea that laboratory science lives up to its self-billing as a controlled, bias- and culture-free environment has been convincingly demolished for some time.[38]

There is of course a level of circularity in arguing that various social (and practical) contexts can explain the fluctuating fortunes of the social context. These arguments are just as historically specific, and deserve some reflection and analysis. Part of the answer is that I seek to analyse what counts as truth in different historical periods. However, this does not answer the question of why I have written an account that focuses partially upon a social context (in the form of intellectual climate in psychology and psychiatry) in order to explain the rise and fall of a socially focused medical category. This might be clarified by explicitly stating my motives for writing this book. I feel deeply uneasy that neurological and neo-liberal explanations (including ideas of human nature, as much as economic policy) are ascendant, to the detriment of socially aware, collective approaches that emphasise the environment and the interpersonal parts of human existence.[39] Market forces, competition and the roll-back of the welfare state – and the acceptance of inequality that this entails – constitute the foundation for mainstream politics in England today. The Labour Party, the founders of the welfare state, are fully signed up (post-2010) to the necessity of 'austerity', and arguably abandoned Keynes in the early-mid 1990s. Their current position seems to be that they would roll back the state just a little more slowly than the Conservative party.[40] (The recent election of 'anti-capitalist' party, Syriza, in Greece might signal a fracturing of the neo-liberal consensus, but it is too early to tell.)

On one level, this history of self-harm is about the organisation of therapeutic approaches and professional practices within health-care systems. In this sense, it has attempted to show how analysis of these areas remains critically important to understanding how and why health epidemics emerge. This account of the establishment and reinforcement of a behavioural pattern also has more intimate consequences. What humans can do, how we experience our emotions and perceive our possibilities – these are fundamentally contextual, situated issues. The turn to social, relational ways of understanding mental health and illness dominate the possibilities for personhood in the middle third of the twentieth century.

The broad point of this book is to show how these possibilities for action or self-experience might come about, and (very briefly) how they might fall away. It is concerned chiefly to reconstruct the intellectual and practical environment where human self-damaging behaviours are chiefly interpreted as communications with a social circle, a cry for assistance and help, in a political climate where there exists considerable consensus about the weight of society's responsibility to provide in a collective manner for the welfare of its citizens (the collective health-care and social-security arrangements referred to as the 'post-war settlement'). It is because that consensus is so thoroughly overturned and almost discarded in the years after 1979 that this book can be seen as politically motivated. The collective aspect of human life is being forgotten in these neurological and neo-liberal reimaginings of human nature.

By using the example of self-damaging behaviour, we take an example that seems to have very little relation to politics (as conventionally conceived), and show how this scientific, clinical object is bound up and implicated with the much larger currents that ebb and flow in the wider culture. These are fuelled on a local level by seemingly mundane practical arrangements, but are no less affected and shaped by the broader intellectual climate. Broad administrative, therapeutic and legal structures interact with local, credible, conceptual and practical labour. This interaction demonstrates the crystallisation and reinforcement of particular intelligible behaviour patterns from infinite possible combinations. This book shows how attempted suicide as communication becomes an available human behaviour pattern at a certain point in history, and how it subsequently becomes displaced. To understand how it is that we act as human, self-conscious beings, we must analyse how the possibilities for comprehensible actions are made. At the same time, we must link these possibilities to the broad political constellations from which the academic humanities seem to be retreating. We must take a position on the ascent of neo-liberalism, as its language of market-friendly research, financial worth and impact continues to take root in academic management. Politically and intellectually, it is a mistake to attempt to explain human complexity, human behaviour and human society through either simplistic market models or flattened biological or evolutionary ones. This is not just because they fail to capture and explain human behaviour in a nuanced and credible way, but also because they are closed systems. They do not allow for their fundamental premises to be questioned or challenged.

Instead, we might affirm the contingency of all explanations, and view with scepticism all claims to unarguable truth. This does not entail political paralysis, but instead invites criticism of what is given or taken for granted in a political (or historical account). At present it is 'given' that any interference with the market creates damaging inefficiencies, that people cutting themselves are responding to internal feelings of tension, that we must 'balance the books' with austerity measures, that our brains hold the key to our selfhood. All of these assertions require constant engagement, criticism and debate.

The point about contingency and scepticism also includes this book. It is written after the 2008 economic crash and bailout, and during the election in Britain of a coalition government of Conservative and Liberal Democrat MPs who are ever more committed to slashing public budgets along with collective responsibility for social problems. In this particular context, it becomes clearer why the text might painstakingly reconstruct a time where the social setting and social interventionism is taken for granted. It establishes a contrast with what is considered so natural in the present (of 2015). Another context concerns funding. The PhD research which forms the basis for this book is funded – made possible – by the Wellcome Trust, a former pharmaceutical company, now a charitable foundation with huge interests in biomedicine and neuroscience as well as in the history of medicine (now more broadly conceived as 'medical humanities'). Indeed, this book is freely accessible on the Internet because of the Trust's generous Open Access (OA) policy to those it funds. My engagement with politics is through the lens of the history of medicine and psychiatry, partially because that falls within the charitable remit of the Trust. This does not mean that I am overplaying or exaggerating the influence of connections between psychiatry, medicine and politics. It simply means that this book (in its present form) would not have been written without the Trust's support. This also needs to be taken into account when weighing the book's contribution and the importance of its emphases and exclusions. I am no more outside my context than the psychiatrists and social workers I study are outside theirs. Money, funding, intellectual fashion – all the things that we willingly forget or skim over when writing academic material – they still matter.

Finally, it is important that reconstructing and analysing the underpinnings of a category based in a social setting is not the same as glorifying or even agreeing with high levels of social intervention. Social work interventionism can lead to horrifying scandals such as that in Cleveland in the North-East of England in the 1980s. Large numbers of children are removed from their families because of allegations (and evidence) of sexual abuse that turn out to be unfounded.[41] So this book is not calling for a 'return to the social' – even if that were possible. It is written instead to call for awareness of the contingency of these organising frameworks. Only by keeping this in our minds can we reach a new consensus where we can weigh our individual and collective responsibilities in a more equitable way. We need to see that the decline in credibility of the social setting, and its replacement by internal self-regulating individuals is among the countless ways in which humans make and remake their worlds (including our ideas of self-damage). The self-evidence of these clinical, psychological and political objects makes them seem natural. This then serves to naturalise the context in which they function – market-based neo-liberalism. If we can see these objects as the result of human actions and human conceptual frameworks, it becomes possible to see that the consequences of the neo-liberal inequalities that assail our society are up for ethical discussion – they are not simply 'human nature' or 'inevitable'. They are, instead, the result of our actions: if we make and accept contexts where inequality is naturalised, then we can also put our efforts into unmaking and refusing these same contexts, and those inequalities.

(cc) (i) BY Except where otherwise noted, this work is licensed under a Creative Commons Attribution 3.0 Unported License. To view a copy of this license, visit http://creativecommons.org/licenses/by/3.0/

Notes

Introduction

1. A rise in self-harm reported in August 2014 called the figures 'alarming', and the idea of self-harm itself as 'deeply distressing'. J. Moorhead. 'Self-harm among Children is on the Rise, But It's Not Just the Victims Who We Need to Support'. http://www.independent.co.uk/voices/comment/selfharm-among-children-is-on-the-rise-but-its-not-just-the-victims-who-we-need-to-support-9662252.html accessed 12 August 2014. For an analysis of media coverage of self-harm in the United States, see W. Bareiss, '"Mauled by a Bear": Narrative Analysis of Self-injury among Adolescents in US News, 2007–2012' *Health: An Interdisciplinary Journal for the Social Study of Health, Illness and Medicine* 18(3) (2014): 279–301

2. S. Gilman, 'From Psychiatric Symptom to Diagnostic Category: Self-harm from the Victorians to DSM-5' *History of Psychiatry* 24 (2013): 149

3. For example the influential: J. Sutton, *Healing the Hurt Within: Understand Self-Injury and Self-Harm, and Heal the Emotional Wounds* 3rd ed. Oxford, How To Books (2007)

4. For example A.R. Favazza, *Bodies Under Siege: Self-Mutilation, Nonsuicidal Self-Injury, and Body Modification in Culture and Psychiatry* 3rd ed. Baltimore, Johns Hopkins University Press (2011)

5. For example Kraus et al., 'Script-driven Imagery of Self-injurious Behavior in Patients with Borderline Personality Disorder: A Pilot FMRI Study' *Acta Psychiatrica Scandinavica* 121(1) (2010): 41–51. I am grateful to Sarah Chaney for making this connection between neurological explanations in general and neurological triggers specifically

6. C. Jacobson and K. Batejan, 'Comprehensive Theoretical Models of Nonsuicidal Self-Injury' in *Oxford Handbook of Suicide and Self-Injury*, M. Nock (ed.) Oxford, Oxford University Press (2014): 312

7. A.R. Favazza, *Bodies Under Siege* 3rd ed. 197, emphasis in original

8. American Psychiatric Association, 'Non-Suicidal Self-Injury' *Diagnostic and Statistical Manual of Mental Disorders 5,* Washington D.C., American Psychiatric Association (2013): 803

9. N. Rose, 'The Death of the Social? Re-figuring the Territory of Government' *Economy and Society* 25(3) (1996): 327–56

10. R. Cooter, 'The End? History-Writing in the Age of Biomedicine (and Before)' in *Writing History in the Age of Biomedicine*, R. Cooter with C. Stein New Haven and London, Yale University Press (2013): 2

11. M. Halewood, *Rethinking the Social Through Durkheim, Marx, Weber and Whitehead* London, Anthem Press (2014): 1

12. I. Hacking, 'Making up People' in *Beyond the Body Proper: Reading the Anthropology of Material Life*, M. Lock and J. Farquhar (eds) Durham, NC, Duke University Press (2007): 150–63

13. B. Brickman, '"Delicate" Cutters: Gendered Self-mutilation and Attractive Flesh in Medical Discourse' *Body and Society* 10(4) (2004): 87–111; C. Millard, 'Making the Cut: the Production of "Self-harm" in Anglo-Saxon psychiatry' *History of the Human Sciences* 26(2) (2012): 126–50

14. J.W. Scott, 'History-writing as Critique' in *Manifestos for History*, K. Jenkins, S. Morgan and A. Munslow (eds) Abingdon, Routledge (2007): 34–5

15. M. Jones, in *Talking about Psychiatry*, G. Wilkinson London, Gaskell (1993): 51; B. Shephard, *A War of Nerves: Soldiers and Psychiatrists in the Twentieth Century* Cambridge, MA, Harvard University Press (2001): 297. Shepherd's information on Gillespie's suicide is an interview with Henry Rollin; there appears to be no other reference in print

16. E. Slater, in *Talking about Psychiatry*, Wilkinson: 8

17. M. Sim, 'Psychological Aspects of Poisoning' in *The Medical Management of Acute Poisoning*, G. Cumming London, Cassell (1961)

18. D.K. Henderson and R.D. Gillespie, *A Text-Book of Psychiatry for Students and Practitioners* 1st ed. London, Humphrey Milford (1927): 133

19. D.K. Henderson and R.D. Gillespie, *A Text-Book of Psychiatry for Students and Practitioners* 6th ed. Oxford, Oxford University Press (1944): vii, 12

20. Henderson and Gillespie, *Text-Book of Psychiatry* 6th ed.: 238

21. Ibid.

22. D.K. Henderson and I.R.C. Batchelor, *Henderson and Gillespie's Textbook of Psychiatry* 9th ed. London, Oxford University Press (1962): 71–4

23. I.R.C. Batchelor, *Henderson and Gillespie's Textbook of Psychiatry for Students and Practitioners* 10th ed. London, Oxford University Press (1969): 72

24. W. Mayer-Gross, M. Roth and E.T.O. Slater, *Clinical Psychiatry* 1st ed. London, Cassell and Co (1954): 198, 211, 264, 502

25. W. Mayer-Gross, M. Roth and E.T.O. Slater, *Clinical Psychiatry* 2nd ed. London, Cassell and Co. (1960): 227, 229

26. E.T.O. Slater and M. Roth, *Clinical Psychiatry* 3rd ed. London, Baillière, Tindall and Cassell (1969): 792

27. For example, N. Kessel, 'The Respectability of Self-poisoning and the Fashion of Survival' *Journal of Psychosomatic Research* 10(1) (1966): 29–36

28. David Aldridge, *Suicide: The Tragedy of Hopelessness* London, Jessica Kingsley (1998): 7; R. Jack, *Women and Attempted Suicide* London, Lawrence Erlbaum (1992): 11

29. M. Sim, *Guide to Psychiatry* 1st ed. Edinburgh and London, E & S Livingstone (1963): 226

30. M. Sim, *Guide to Psychiatry* 2nd ed. Edinburgh and London, E & S Livingstone (1968): 610

31. B. Brickman, '"Delicate" Cutters'; Millard, 'Making the Cut'; M.A. Simpson 'Self Mutilation and Suicide' in *Suicidology: Contemporary Developments*, E.S. Shneidman (ed.) New York, Grune and Stratton (1976): 286–315

32. M. Sim, *Guide to Psychiatry* 3rd ed. (With a chapter on legal aspects of psychiatry in the United States of America by John Donnelly) Edinburgh and London, Churchill Livingstone (1974): 704–5

33. M. Sim, *Guide to Psychiatry* 4th ed. (With a chapter on legal aspects of psychiatry in the United States of America by John Donnelly) Edinburgh and London, Churchill Livingstone (1981): 432

34. N. Kreitman, 'Can Suicide and Parasuicide Be Prevented?' *Journal of the Royal Society of Medicine* 82(11) (1989): 650, 648
35. N. Kreitman, A.E. Philip, S. Greer and C.R. Bagley, 'Parasuicide' *British Journal of Psychiatry* 115 (1969): 747
36. See N. Kreitman (ed.), *Parasuicide* London, John Wiley & Sons (1977): 1; W.H. Trethowan, 'Suicide and Attempted Suicide' *British Medical Journal* 2, 6185 (1979): 320
37. E. Stengel, 'Enquiries into attempted suicide (Abridged)' *Proceedings of the Royal Society of Medicine* 45 (1952): 613
38. O. Anderson, 'Prevention of Suicide and Parasuicide: What Can We Learn from History?' *Journal of the Royal Society of Medicine* 82 (11) (1989): 642
39. Anderson, 'Suicide and Parasuicide': 642
40. Kreitman et al., 'Parasuicide': 747, emphasis in original.
41. Sarah Chaney deftly avoids such a loss when investigating the contextually specific meanings of 'self-mutilation' in Victorian psychology. S. Chaney, '"A Hideous Torture on Himself": Madness and Self-mutilation in Victorian Literature' *Journal of Medical Humanities* 32 (2011): 279–89
42. Anderson, 'Suicide and Parasuicide': 642. See also: O. Anderson, *Suicide in Victorian and Edwardian England* Oxford, Clarendon (1987): 263–417
43. D. Tantam and N. Huband, *Understanding Repeated Self-Injury: A Multidisciplinary Approach* Basingstoke, Palgrave Macmillan (2009): 4
44. Sutton, *Healing the Hurt Within* (2007): 105–6
45. For an early example of this, see B. Walsh and P. Rosen, *Self-mutilation: Theory, Research and Treatment* Guildford, Guildford Press (1988): vii; see also J. Hyman, *Women Living With Self-Injury* Philadelphia, Temple University Press (1999): 193
46. K. Hawton, 'Deliberate Self-poisoning and Self-injury in the Psychiatric Hospital' *British Journal of Medical Psychology* 51(3) (1978): 257–8 (see Chapter 5)
47. See Millard 'Making the cut': 127; Sutton *Healing the Hurt Within*: 116–23
48. A.R. Favazza, *Bodies under Siege: Self-Mutilation and Body Modification in Culture and Psychiatry* 2nd ed. Baltimore, Johns Hopkins University Press (1996): 2, 11–16
49. J. Sutton, *Healing the Hurt Within* 108
50. Dunlop is most famous for the 'Dunlop Committee' on drug safety established in 1964 following the Thalidomide disaster (see Chapter 4). D.M. Dunlop, 'Foreword' to *Treatment of Common Acute Poisonings*, H.J.S. Matthew and A.A.H. Lawson Edinburgh, E. & S. Livingstone (1967): iv
51. J.W. Scott, *Gender and the Politics of History* rev. ed. New York, Columbia University Press (1999): 7
52. R. Leys, 'How Did Fear Become a Scientific Object and What Kind of Object Is It?' *Representations* 110 (2010): 66–104
53. I. Hacking, *Rewriting the Soul* Princeton, NJ, Princeton University Press (1995): 234–67
54. Hacking, *Rewriting*: 242, 248
55. K. McMillan, 'Under a redescription' *History of the Human Sciences* 16(2) (2003): 136
56. A. Young, *The Harmony of Illusions: Inventing Post-Traumatic Stress Disorder* Princeton, NJ, Princeton University Press (1995): 5

57. A. Wilson, 'On the History of Disease Concepts: The Case of Pleurisy' *History of Science* 38 (2000): 273
58. A. Shepherd and D. Wright, 'Madness, Suicide and the Victorian Asylum: Attempted Self-murder in the Age of Non-restraint' *Medical History* 46(2) (2002): 179, 193
59. Shepherd and Wright, 'Madness, Suicide and the Victorian Asylum': 194–6
60. Å. Jansson, 'From Statistics to Diagnostics: Medical Certificates, Melancholia, and "Suicidal Propensities" in Victorian Medicine' *Journal of Social History* 46(3) (2013): 716–31
61. S. Chaney, 'Suicide, Mental Illness and the Asylum: The Case of Bethlem Royal Hospital 1845–1875' MA Dissertation, University College London (2009)
62. F. Hopkins, 'Attempted Suicide: An Investigation' *Journal of Mental Science* 83 (1937): 90
63. E. Stengel, 'Attempted Suicide and the Law' *Medico-Legal Journal* 27 (1959): 115
64. M. Foucault, *The Essential Works of Michel Foucault, 1954–1984*. Vol. 2, *Aesthetics* London, Penguin (2000): 463
65. For example R. Porter, *The Greatest Benefit to Mankind: A Medical History of Humanity from Antiquity to the Present* London, Fontana Press (1999): 493–4
66. G.G. Lloyd and R.A. Mayou, 'Liaison Psychiatry or Psychological Medicine?' *British Journal of Psychiatry* 183 (2003): 5
67. Department of Health 'The Configuration of Mental Health Services' (2007) http://www.dh.gov.uk/en/Publicationsandstatistics/Publications/PublicationsPolicyAndGuidance/Browsable/DH_4897913 accessed 13 July 2012
68. For example, W.S. Maclay, 'After the Mental Health Act: An Appraisal of English Psychiatry' *Mental Hospitals* 14 (1963): 100
69. See for example, P. Barham, *Closing the Asylum: The Mental Patient in Modern Society* Harmondsworth, Penguin (1992); P. Brown, *The Transfer of Care: Psychiatric Deinstitutionalization and Its Aftermath* Boston, Routledge & Kegan Paul (1985)
70. For a recent example, see A. Lovell and E. Susser (eds) *International Journal of Epidemiology* Special Issue on the History of Psychiatric Epidemiology (2014)
71. M. Jackson, *The Age of Stress: Science and the Search for Stability* Cambridge, Cambridge University Press (2013): 171
72. H. Selye, *Stress of Life* London, McGraw-Hill (1956); W.B. Cannon, *Bodily Changes in Pain, Hunger, Fear and Rage* New York, D. Appleton & Co. (1929)
73. Jackson, *Age of Stress:* 180
74. T.H Holmes and R.H. Rahe, 'The Social Readjustment Rating Scale' *Journal of Psychosomatic Research* 11(2) (1967): 213–18; G.W. Brown and T. Harris, *Social Origins of Depression: A Study of Psychiatric Disorder in Women* London, Tavistock Publications (1978)
75. Young, *Harmony of Illusions*
76. R. Hayward, 'Sadness in Camberwell: Imagining Stress and Constructing History in Post War Britain [Draft]' paper given in Washington, D.C. (2010): 2
77. M. Parascandola, 'Epidemiology in Transition: Tobacco and Lung Cancer in the 1950s' in *Body Counts: Medical Quantification in Historical and Sociological Perspective*, G. Jorland, A. Opinel, and G. Weisz (eds) Ithaca, McGill-Queens University Press (2005): 226

78. A.D. Gardner 'Debauchery of Honest Words' *British Medical Journal* 1, 4760 (1952): 715
79. M. Shepherd, 'Epidemiology and Clinical Psychaitry' *British Journal of Psychiatry* 133 (1978): 289
80. Letter from G.M. Carstairs to H. Himsworth (MRC Secretary) dated 01 July 1959, The National Archives (TNA) FD 7/1043: Proposed Unit for research into the Epidemiology of Mental Diseases, University College Hospital, London, 1958–9
81. G.M. Carstairs, 'Memorandum to Council and CRB': 'Proposed Unit for Research on the Epidemiology of Mental Disorders' 1 TNA FD 7/1043; D.D. Reid, *Epidemiological Methods in the Study of Mental Disorders* Geneva: World Health Organization (1960): 7
82. A. Lewis, 'Social Aspects of Psychiatry: Part I' *Edinburgh Medical Journal* 58(5) (1951): 215
83. D. Armstrong, *Political Anatomy of the Body: Medical Knowledge in Britain in the Twentieth Century* Cambridge, Cambridge University Press (1983): 102
84. Armstrong, *Political Anatomy:* 114
85. N. Kessel, 'Self-Poisoning – Part II' *British Medical Journal* 2, 5474 (1965): 1336
86. See also C. Millard 'Stress Attempted Suicide and The Social' in *Stress in Post-War Britain* M. Jackson (ed.) Studies for the Society for the Social History of Medicine, Pickering and Chatto (forthcoming)
87. J. Ruesch and G. Bateson, *Communication: The Social Matrix of Psychiatry* New York, W.W. Norton (1951): 3, 50, 78–9
88. T.S. Szasz, *The Myth of Mental Illness: Foundations of a Theory of Personal Conduct* London, Paladin ([1961] 1972): 25, 128–52
89. I. Hacking, *Historical Ontology* London, Harvard University Press (2002): 107
90. Scott, *Gender and the Politics of History:* 10–11
91. N. Rose, *Politics of Life Itself: Biomedicine, Power and Subjectivity in the Twenty-First Century* Princeton, NJ, Princeton University Press (2007): 5
92. M. Foucault, *Politics, Philosophy, Culture: Interviews and Other Writings, 1977–1984* London, Routledge (1988): 154

1 Early Twentieth-Century Self-Harm: Cut Throats, General and Mental Medicine

1. Report by P.C. Albert E. Turner dated 29 June 1914 [in error]: CRIMINAL; POLICE: Removal of attempted suicides to hospital: police liability to guard patient to prevent another attempt, 1907–50 TNA: HO 45/24439
2. Letter from Inspector Fredk. Peck, to Supt. Page, Deputy Chief Constable of East Suffolk dated 17 July 1914 TNA: HO 45/24439
3. Letter from Peck to Page 17 July 1914 TNA: HO 45/24439. Ashby has committed the common law misdemeanour of 'attempted suicide', not the felony of 'attempted murder'.
4. 'East Suffolk Constabulary General Orders' dated 05 March 1902 TNA: PRO: HO 45/24439
5. Draft Reply to the Chief Constable of East Suffolk dated 05 August 14 TNA: PRO HO45/24439

6. S. Chaney, '"A Hideous Torture on Himself" Madness and Self-Mutilation in Victorian Literature' *Journal of Medical Humanities* 32 (2011): 280–1
7. It is unclear when the practice of police watching emerges, but it is probable that it comes to renewed prominence in the mid-nineteenth century, when 'attempted suicide' becomes a common-law offence, what Olive Anderson calls the 'new offence'. O. Anderson, *Suicide in Victorian and Edwardian England* Oxford, Clarendon (1987): 263–417
8. See Chapter 2.
9. Minute by Norman Kendal (New Scotland Yard) 11 July 1930 TNA: HO 45/24439
10. M. Gorsky and J. Mohan, 'London's Voluntary Hospitals in the Interwar Period: Growth, Transformation, or Crisis?' *Nonprofit and Voluntary Sector Quarterly* 30(2) (2001): 248; G. Rivett, 'The Voluntary Hospitals' in *Development of the London Hospital System 1823–1992* online at: http://www.nhshistory.net/voluntary_hospitals.htm accessed 30 January 2015
11. C.A.H. Watts, *Depressive Disorders in the Community* Bristol, John Wright & Sons (1966): 1
12. Letter from Home Office to Deputy Chief Constable of Wiltshire dated 16 August 1915 TNA: HO 45/24439
13. Letter from W. Bryan Forward (Lowestoft and North Suffolk Hospital) to Home Office, dated 06 June 1914 TNA: HO45/24439
14. Letter from the Home Office to Chief Constable of East Suffolk dated 05 August 1914 TNA: PRO: HO 45/24439
15. Letter from Commissioner of Police for the Metropolis to the Secretary of the British Hospitals Association dated 26 January 1923 TNA: HO 45/24439
16. Minute on file by H.B. Simpson (draft letter to F. Caldwell) dated 26 October 1920 TNA: HO 45/24439
17. Draft Letter from Home Office to Huntingdon Constabulary dated 11 July 1907 TNA: HO 45/24439
18. Rivett, 'The Voluntary Hospitals'
19. Chief Constable of Liverpool to Home Office dated 21 October 1920 TNA: HO 45/24439
20. Home Office to Chief Constable of Liverpool dated 1 November 1920 TNA: HO 45/24439
21. 'Man with cut throat' *The Reporter* 04 February 1922 TNA: HO 45/24439
22. H.L. Freeman, 'Psychiatry in the National Health Service 1948–1998' *British Journal of Psychiatry* 175 (1999): 3
23. R. Mayou, 'The History of General Hospital Psychiatry' *British Journal of Psychiatry* 155 (1989): 768.
24. Letter from F. Oliver (District Infirmary) to H.H. Daley (Lake Hospital) dated 01 February 1922 TNA: HO 45/24439; see also Staffordshire Constabulary Memo dated 30 December 1910 TNA: HO 45/24439
25. Letter from Oliver to Daley dated 01 February 1922 TNA: HO 45/24439
26. 'Man with cut throat' *The Reporter* 04 February 1922 TNA: HO 45/24439
27. 'Hospitals and the Home Office' *Manchester Guardian* 29 April 1922 TNA: HO 45/24439
28. East Suffolk Constabulary General Orders dated 05 March 1902 TNA: HO 45/24439

29. Standing Order no. 7 from Staffordshire Constabulary 'Cases of Persons who have attempted to Commit Suicide' dated November 1904 TNA: HO 45/24439

30. Letter from Sir Edward R. Henry (Metropolitan Police Commissioner) to Home Office dated 27 July 1916 in 'Police Supervision of AS in Hospitals' Attempted suicide: police supervision of offenders in hospital, 1916–1951 TNA: MEPO 3/2436

31. Mayou, 'General Hospital Psychiatry': 768

32. D.R. Benady and J. Denham, 'Development of an Early Treatment Unit from an Observation Ward' *British Medical Journal* 2, 5372 (1963): 1569

33. J. Pickstone, 'Psychiatry in District General Hospitals' in *Medical Innovations in Historical Perspective* J. Pickstone (ed.) (1992): 198

34. D. Armstrong, *Political Anatomy of the Body: Medical Knowledge in Britain in the Twentieth Century* Cambridge, Cambridge University Press (1983): 73

35. Lewis and Calder, 'General Report on Observation Wards' Maudsley Hospital, Denmark Hill, London SE5, 1923–1948 TNA: MH 95/32; see also Mayou's slightly different assessment: Mayou, 'General Hospital Psychiatry': 768

36. M.D. Eilenberg, M.J. Pritchard and P.B. Whatmore, 'A 12-Month Survey of Observation Ward Practice' *British Journal of Preventive and Social Medicine* 16 (1962): 22

37. A.D., 'Obituary Notices: Hon. W.S. Maclay' *British Medical Journal* 1, 5392 (1964): 1258

38. W.S. Maclay, 'After the Mental Health Act: An Appraisal of English Psychiatry' *Mental Hospitals* 14 (1963): 100

39. W.S. Maclay, 'Trends in the British Mental Health Service' in *Trends in the Mental Health Services: A Symposium of Original and Reprinted Papers* H.L. Freeman and W.A.J. Farndale (eds) Oxford, Pergamon Press (1963)

40. B. Wootton, *Social Science and Social Pathology* London, George Allen & Unwin, (1959) 208

41. E.U.H. Pentreath and E.C. Dax, 'Mental Observation Wards: A Discussion of Their Work and Its Objects' *Journal of Mental Science* 83 (1937): 351–2

42. A. Lewis and F.H.M. Calder, 'A General Report on the Observation Wards Administered by the London County Council' 10 (1938) TNA: MH 95/32

43. F. Hopkins, 'Admissions to Mental Observation Wards During War' *British Medical Journal* 1, 4289 (1943): 358; see also F. Hopkins 'Attempted Suicide: An Investigation' *Journal of Mental Science* 83 (1937): 71; and below

44. Lewis and Calder, 'General Report on the Observation Wards' 21 TNA: MH 95/32

45. Ibid. 3–5 TNA: MH 95/32

46. Pentreath and Dax, 'Mental Observation Wards': 362

47. Quoted in Pentreath and Dax, 'Mental Observation Wards': 348–9

48. I. Skottowe, 'Discussion: Observation Units' *Proceedings of the Royal Society of Medicine* 33 (1940): 732, emphasis in the original.

49. E.W. Dunkley and E. Lewis, 'North Wing: A Psychiatric Unit in a General Hospital' *Lancet* 281, 7273 (1963): 156

50. See for example, J. Bourke, *Dismembering the Male: Men's Bodies, Britain and the Great War* London, Reaktion Books (1996); B. Shephard, *A War of Nerves: Soldiers and Psychiatrists in the Twentieth Century* Cambridge, MA, Harvard University Press (2001)

51. T.H. Goodwin, 'The Casualty Clearing Stations' *Journal of the American Medical Association* 69(8) (1917): 636
52. Armstrong, *Political Anatomy:* 7–8
53. Pentreath and Dax, 'Mental Observation Wards': 362
54. Lewis and Calder 'General Report on the Observation Wards' 6, 16 TNA MH 95/32. Pentreath's and Dax's article is quoted at length on 14
55. The exact timing of these visits is unclear; that they were regular is not disputed. See Mayou, 'General Hospital Psychiatry': 768; G. Wilkinson, *Talking About Psychiatry* London, Gaskell, (1993): 138
56. Pentreath and Dax, 'Mental Observation Wards': 363
57. Lewis and Calder, 'General Report on the Observation Wards' 21 TNA MH 95/32
58. Quoted in R.P. Snaith and S. Jacobson, 'The Observation Ward and the Psychiatric Emergency' *British Journal of Psychiatry* 111 (1965): 18
59. Pentreath and Dax, 'Mental Observation Wards': 347, 349
60. J. Marshall, 'Mental Health Services' *British Medical Journal* 2 (4902) (1954): 1484
61. M.B. Hall and F. Hopkins, 'Parental Loss and Child Guidance' *Archives of Disease in Childhood* 11(64) (1936): 187–94
62. I. Leveson, 'Evolution of Psychiatry in a Clinician's Lifetime' *Transactions and Report/Liverpool Medical Institution* (1968): 23–5
63. Hopkins, 'Attempted Suicide': 71
64. Letter from Chief Constable of Liverpool to Home Office dated 21 October 1920 CRIMINAL; POLICE: Removal of attempted suicides to hospital: police liability to guard patient to prevent another attempt, 1907–1950 TNA HO 45/24439
65. Hopkins, 'Admissions': 358
66. Hopkins, 'Attempted Suicide': 71
67. Ibid. 93
68. Ibid. 71–2, 84–5
69. Ibid. 85
70. Ibid. 76–7
71. Ibid. 78
72. Ibid. 91
73. Ibid. 85–6
74. Ibid. 90
75. Hopkins and Hall characterise the 'disturbed' or 'broken home' as 'a medley of facts and circumstances, decisive in their origin, but diffuse in their manifestations and extensive in their effects'. Hall and Hopkins, 'Parental Loss': 194

2 Communicative Self-Harm: War, NHS and Social Work

1. D.K. Henderson and R.D. Gillespie, *A Text-Book of Psychiatry for Students and Practitioners*, 6th ed. Oxford, Oxford University Press (1944): vii
2. J. Toms, *Mental Hygiene and Psychiatry in Modern Britain* Basingstoke, Palgrave Macmillan (2013): 32–3

3. Department of Health, 'Your Very Good Health' (1948) online at: http://www. nhs.uk/Livewell/NHS60/Pages/VideointroducingthenewNHS.aspx   accessed 11 July 2012
4. Samaritans, 'Samaritans History' online at: http://www.samaritans.org/ about_samaritans/governance_and_history/samaritans_history.aspx accessed 11 July 2012
5. J.R. Rees, *The Shaping of Psychiatry by War* London, Chapman & Hall (1945): 10, 33
6. T. Harrison, *Bion, Rickman, Foulkes and the Northfield Experiments: Advancing on a Different Front* London, Jessica Kingsley (2000): 27–35, 73
7. J.M. Mackintosh, *The War and Mental Health in England* New York, Commonwealth Fund (1944): 62
8. Mackintosh, *War and Mental Health:* 63
9. Harrison, *Advancing on a Different Front:* 25, 76
10. Ibid. 73–4
11. R. Mayou, 'The History of General Hospital Psychiatry' *British Journal of Psychiatry* 155 (1989): 770
12. W.S. Maclay, 'Trends in the British Mental Health Service' in *Trends in the Mental Health Services: A Symposium of Original and Reprinted Papers* H.L. Freeman and W.A.J. Farndale (eds) Oxford, Pergamon Press (1963): 4; J.K. Wing, *Reasoning About Madness* Oxford, Oxford University Press (1978): 199
13. J. Carson and E.H. Kitching, 'Psychiatric Beds in a General Ward: A Year's Experience' *Lancet* 1, 6559 (1949): 833
14. J. Pickstone, 'Psychiatry in District General Hospitals' in *Medical Innovations in Historical Perspective* J. Pickstone (ed.) (1992): 191–2
15. G. Wilkinson, *Talking About Psychiatry* London, Gaskell, (1993): 147
16. H.R. Guly, *A History of Accident and Emergency Medicine, 1948–2004* Basingstoke, Palgrave Macmillan (2005): 4, xii
17. One of a huge number of examples: D. Campbell 'Overcrowded hospitals "killed 500" last year, claims top A&E doctor' *Observer* (24 January 2015) online at: http://www.theguardian.com/uk-news/2015/jan/24/over-crowded-hospitals-deaths accessed 26 January 2015
18. T.G. Lowden, 'The Casualty Department. I. The Work and the Staff' *Lancet* 270, 6929 (1956): 955
19. T.G. Lowden, *The Casualty Department* London, E. & S. Livingstone (1955): 254
20. Lowden, *The Casualty Department:* 254
21. J. Carson and E.H. Kitching, 'Psychiatric Beds in a General Ward: A Year's Experience' *Lancet* 1, 6559 (1949): 833
22. I.R.C. Batchelor, 'Attempted Suicide' *British Medical Journal* 1, 4913 (1955): 595; see also R.A.J. Asher, 'Arrangements for the Mentally Ill' *Lancet* 268, 6955 (1956): 1266; N. Kessel and G. Grossman, 'Suicide in Alcoholics' *British Medical Journal* 2, 5268 (1961): 1672
23. P.G. Aungle, 'Sir Ivor Batchelor' *Psychiatric Bulletin* 29 (2005): 439
24. S. Crown, 'Post-War Maudsley Personalities' *Psychiatric Bulletin* 12(7) (1988): 264
25. D. Odlum, P. Epps, I.R.C. Batchelor and I.M.H. McAdam, 'Discussion on the Legal Aspects of Suicidal Acts' *Proceedings of the Royal Society of Medicine* 51(4) (1958): 297–303

26. E. Stengel, 'Attempted Suicide and the Law' *Medico-Legal Journal* 27(1959): 118

27. I.R.C. Batchelor and M.B. Napier, 'The Sequelae and Short-Term Prognosis of Attempted Suicide; the Results of a One-Year Follow-up of 200 Cases' *Journal of Neurology, Neurosurgery & Psychiatry* 17(4) (1954): 265–6

28. Letter from A.B. Hume to F.L.T. Graham-Harrison dated 16 June 1958 Law in relation to attempted suicide: general correspondence, 1958–1961 TNA: MH 137/383

29. Letter from P.A. Cox to J.D.J. Havard dated 16 January 1962 Law in relation to attempted suicide: proposed amendments, 1961–1962 TNA: MH 137/384. There is at least one similar arrangement in England, in Sheffield from 1951 (see Chapter 3).

30. N. Kessel, 'Self-Poisoning – Part I' *British Medical Journal* 2, 5473 (1965): 1265

31. D. Tait, 'Norman Kreitman in Conversation with David Tait' *Psychiatric Bulletin* 19 (1995): 298

32. The memo is unattributed, but almost certainly written by J.K. Slater: 'Ward 3: A Revaluation for the Ward for Incidental Delirium' (October 1962) Proposed Regional Poisoning Treatment Centre; revaluation of functions of Ward 3; poisoning among children, 1962 at Lothian Health Services Archive (LHSA), University of Edinburgh: LHB 1/59/7: 4. Graduates include G.M. Carstairs, head of Kessel's and Kreitman's research unit.

33. [Slater], 'Ward 3: A Revaluation' LHSA: LHB 1/59/7 1

34. Ibid. 3

35. Ibid. 3

36. Ibid. 3–4.

37. Ibid. 4

38. I. Hacking, *The Taming of Chance* Cambridge, Cambridge University Press (1990); 65. Hacking is talking about the claims by French alienist J.E. Esquirol to cast 'suicide as a kind of madness'.

39. N. Kessel, 'Attempted Suicide' *Medical World* 97 (1962): 314

40. E. Stengel, 'Attempted Suicide: Its Management in the General Hospital' *Lancet* 1, 7275 (1963): 234

41. E.F. Catford, *The Royal Infirmary of Edinburgh 1929–1979* Edinburgh, Scottish Academic Press (1984): 159

42. V. Long, '"Often There Is a Good Deal to Be Done, but Socially Rather Than Medically": The Psychiatric Social Worker as Social Therapist, 1945–70' *Medical History* 55(2) (2011): 225

43. N.K. Hunnybun, 'Psychiatric Social Work' in *Social Case-Work in Great Britain* C. Morris (ed.) London: Faber and Faber (1950): 102

44. Toms, *Mental Hygiene and Psychiatry in Modern Britain*: 30

45. J. Stewart, '"I Thought You Would Want to Come and See His Home": Child Guidance and Psychiatric Social Work in Inter-War Britain' in *Health and the Modern Home* M. Jackson (ed.) London, Routledge (2007): 111–12; Hunnybun, 'Psychiatric Social Work' 105; E. Heimler, *Mental Illness and Social Work* Harmondsworth, Penguin (1967): 44–5

46. Hunnybun, 'Psychiatric Social Work': 103–4

47. Rees, *Shaping of Psychiatry*: 32

48. Toms, *Mental Hygiene*: 86

49. J. Bowlby, *Maternal Care and Mental Health* Geneva, World Health Organization (1951)
50. A. Storr, 'Bowlby, (Edward) John Mostyn (1907–1990)' in *Oxford Dictionary of National Biography* Oxford, Oxford University Press (2004)
51. Long, 'Good Deal to Be Done': 226
52. E. Irvine, 'Psychiatric Social Work' in E. Younghusband, *Social Work in Britain, 1950–1975: A Follow-Up Study* London, Allen & Unwin (1978): 179.
53. F. Post, 'Mental Breakdown in Old Age' *British Medical Journal* 1, 4704 (1951): 436–40; F. Post and J. Wardle, 'Family Neurosis and Family Psychosis' *British Journal of Psychiatry* 108(1962): 147
54. Irvine, 'Psychiatric Social Work': 178, 192. Irvine is talking specifically about 1950s mental hospitals.
55. E. Younghusband, *Social Work in Britain, 1950–1975: A Follow-up Study* London, Allen and Unwin (1978): 165
56. Toms, *Mental Hygiene:* 90
57. M. Jones, *Social Psychiatry in Practice: The Idea of the Therapeutic Community* Harmondsworth, Penguin (1968): 16–17
58. Rees, *Shaping of Psychiatry:* 32
59. Bion, *Experiences in Groups:* 13
60. T. Main, quoted in Toms, *Mental Hygiene:* 101
61. Toms, *Mental Hygiene:* 115
62. Mackintosh, *War and Mental Health:* 59
63. For an interesting account of the turn to the 'social' in the United States, see M.E. Staub, *Madness Is Civilization: When the Diagnosis Was Social, 1948–1980* London, University of Chicago Press (2011)
64. G.W. Brown, E.M. Monck, G.M. Carstairs and J.K. Wing, 'Influence of Family Life on the Course of Schizophrenic Illness' *British Journal of Preventive and Social Medicine* 16(2) (1962): 55–68
65. N. Rose, *Governing the Soul : The Shaping of the Private Self* London, Free Association Press (1999): 171, 175
66. M. Thomson, *Psychological Subjects: Identity, Culture, and Health in Twentieth-Century Britain* Oxford, Oxford University Press (2006): 269. Thomson is describing the *aspirations* for a new, enlightened, psychologised sense of selfhood.
67. G. Eghigian, A. Killen, and C. Leuenberger, 'The Self as Project: Politics and the Human Sciences in the Twentieth Century' *Osiris* 22(1) (2007): 22
68. Rose, *Governing the Soul:* 155–81, 205–13
69. Toms, *Mental Hygiene*
70. E.L. Younghusband, *Social Work in Britain: A Supplementary Report on the Employment and Training of Social Workers* Dunfermline, Carnegie United Kingdom Trust (1951): 81
71. Ministry of Health, 'Report of the Working Party on Social Workers in the Local Authority Health and Welfare Services (Younghusband Report)' London, HMSO (1959)
72. L. Faithfull, 'Younghusband, Dame Eileen Louise (1902–1981)' in *Oxford Dictionary of National Biography* Oxford, Oxford University Press (2004)
73. R.M. Titmuss, 'Community Care: Fact or Fiction?,' in *Commitment to Welfare*, R.M. Titmuss (ed.) London, George Allen and Unwin ([1961] 1968): 105
74. Rose, *Governing the Soul:* 208

75. A. Lewis, 'Social Aspects of Psychiatry: Part I' *Edinburgh Medical Journal* 58(5) (1951): 215
76. Younghusband, *Social Work in Britain:* 165
77. E.U.H. Pentreath and E.C. Dax, 'Mental Observation Wards: A Discussion of Their Work and Its Objects' *Journal of Mental Science* 83 (1937): 354, 364
78. Despite Hopkins positing domestic stress as a precipitating factor for attempted suicide and publishing his observation ward study in the same year, he does not mention a social worker (psychiatric or otherwise). He does, however, have an interest in child guidance.
79. E.N. Butler, 'Observation Units' *Proceedings of the Royal Society of Medicine* 33 (1940): 726.
80. Hunnybun, 'Psychiatric Social Work': 117.
81. I.R.C. Batchelor and M.B. Napier, 'The Sequelae and Short-Term Prognosis of Attempted Suicide; the Results of a One-Year Follow-up of 200 Cases' *Journal of Neurology, Neurosurgery & Psychiatry* 17(4) (1954): 261; I.R.C Batchelor and M.B. Napier, 'Attempted Suicide in Old Age' *British Medical Journal* 2, 4847 (1953): 1187; I.R.C. Batchelor, 'Alcoholism and Attempted Suicide' *Journal of Mental Science* 100, 419 (1954): 461
82. Batchelor and Napier, 'Sequelae': 261
83. Stewart, 'See His Home': 115–17
84. Ibid. 118
85. Bridget Yapp, quoted in Stewart, 'See His Home': 118. In 1978, Irvine claims that this text is 'the only British book on the subject' and that it is 'unfortunate that no other books had appeared.' Irvine 'Psychiatric Social Work': 183
86. E.T.O. Slater and M. Woodside, *Patterns of Marriage: A Study of Marriage Relationships in the Urban Working Classes* London, Cassell (1951) 14. See Chapter 4 for more on the influence of marriage guidance and PSWs in attempted-suicide studies.
87. Stewart, 'See His Home': 118
88. Hunnybun, 'Psychiatric Social Work': 101
89. Irvine, 'Psychiatric Social Work': 178
90. Bowlby quoted in Toms, *Mental Hygiene* 114
91. Toms, *Mental Hygiene* 114
92. Batchelor and Napier, 'Sequelae': 265
93. I.R.C. Batchelor, 'Management and Prognosis of Suicidal Attempts in Old Age' *Geriatrics* 10(6) (1955): 292
94. Batchelor and Napier, 'Sequelae': 266
95. Ibid. 264
96. Ibid. 264
97. Ibid. 264, 266
98. I.R.C. Batchelor and M.B. Napier, 'Broken Homes and Attempted Suicide' *British Journal of Delinquency* 4 (1953): 99
99. Batchelor and Napier, 'Broken Homes': 99–100
100. Bowlby, 'Maternal Care and Mental Health': 12
101. Ibid. 105–6
102. Ibid. 101
103. Ibid. 104
104. Ibid. 103–4

105. I.R.C. Batchelor, 'Psychopathic States and Attempted Suicide' *British Medical Journal* 1, 4875 (1954): 1343
106. Batchelor and Napier, 'Broken Homes': 103–4
107. Batchelor, 'Alcoholism': 453
108. Batchelor, 'Psychopathic States': 1343
109. Batchelor and Napier, 'Broken Homes': 104
110. Ibid. 105
111. J.W. Scott, *Gender and the Politics of History* rev. ed. New York, Columbia University Press (1999) 115
112. Batchelor and Napier, 'Broken Homes': 107–8
113. I.R.C. Batchelor, 'Repeated Suicidal Attempts' *British Journal of Medical Psychology* 27(3) (1954): 161
114. They also argue that 'those rendered most vulnerable by their early experiences may tend to break down quickly' Batchelor and Napier, 'Broken Homes': 102, 106
115. Bowlby, 'Maternal Care and Mental Health': 13–14
116. Rose, *Governing the Soul* 170
117. See M. Vicedo, 'The Social Nature of the Mother's Tie to Her Child: John Bowlby's Theory of Attachment in Post-War America' *British Journal for the History of Science* 44(3) (2011): 420
118. Bowlby, 'Maternal Care and Mental Health': 26
119. Mayou, 'General Hospital Psychiatry': 774
120. Wilkinson, *Talking about Psychiatry:* 28
121. W.H. Trethowan, 'Suicide and Attempted Suicide' *British Medical Journal* 2, 6185 (1979): 320
122. Trethowan, 'Suicide and Attempted Suicide': 320
123. Stengel's biographical details are taken from four sources: P. Weindling, 'Alien Psychiatrists: The British Assimilation of Psychiatric Refugees' in *International Relations in Psychiatry: Britain, Germany, and the United States to World War II* V. Roelcke, P. Weindling and L. Westwood (eds) Rochester, University of Rochester Press (2010): 218–36; U.H. Peters, 'The Emigration of German Psychiatrists to Britain' in *150 Years of Psychiatry: The Aftermath* H.L. Freeman and G.E. Berrios (eds) London, Athlone (1996): 565–80; F.A. Jenner, 'Stengel, Erwin (1902–1973)' in *Oxford Dictionary of National Biography* Oxford, Oxford University Press (2004); M. Shepherd, 'The Impact of Germanic Refugees on Twentieth-Century British Psychiatry' *Social History of Medicine* 22(3) (2009): 461–9
124. E. Stengel, 'On the Aetiology of the Fugue States' *Journal of Mental Science* 87 (1941): 572–99; E. Stengel, 'Further Studies on Pathological Wandering (Fugues with the Impulse to Wander)' *Journal of Mental Science* 89 (1943): 224–41
125. E. Stengel, 'Suicide' in *Recent Progress in Psychiatry* G.W.T.H. Fleming, A. Walk and P.K. McCowan (eds) London, J. & A. Churchill Ltd (1950): 691–703
126. Anonymous, 'Lost Hospitals of London' online at: http://ezitis.myzen.co.uk/stfrancis.html accessed 21 August 2012
127. Wilkinson, *Talking*, 30, 231; Post, 'Mental Breakdown in Old Age'
128. E. Stengel, N.G. Cook and I.S. Kreeger, *Attempted Suicide: Its Social Significance and Effects* London, Chapman and Hall (1958): 22, emphasis in original.
129. Stengel, Cook and Kreeger, *Attempted Suicide:* 43, 84, 105

130. V. Norris, *Mental Illness in London* London, Chapman & Hall (1959): 91
131. D.F. Early, 'The Changing Use of the Observation Ward' *Public Health* 76(5) (1962): 262
132. J. Marshall, 'Mental Health Services' *British Medical Journal* 2, 4902 (1954): 1484; J.B.S. Lewis, 'Mental Health Services' *British Medical Journal* 2, 4900 (1954): 1354–5
133. Norris, *Mental Illness* 234; see also, M.D. Eilenberg and P.B. Whatmore, 'Police Admissions to a Mental Observation Ward' *Medicine, Science, and the Law* 2 (1961): 96–100
134. Wilkinson, *Talking about Psychiatry* 15; this story also appears in Mayou, 'General Hospital Psychiatry': 774
135. Stengel, Cook and Kreeger, *Attempted Suicide:* 96
136. J.J. Fleminger and B.L. Mallett, 'Psychiatric Referrals from Medical and Surgical Wards' *British Journal of Psychiatry* 108 (1962): 189
137. Stengel, Cook and Kreeger, *Attempted Suicide:* 47, 93–4
138. Ibid. 40
139. Ibid. 41
140. Ibid. 39
141. Stengel, 'Enquiries into Attempted Suicide (Abridged)': 616
142. There are also categories less relevant for the new 'cry for help object' such as 'Suicidal attempt followed by permanent institutionalisation' and 'suicidal attempt followed by death soon' Stengel, Cook and Kreeger, *Attempted Suicide:* 52, 55–7
143. Of the nine cases under the heading, two attempts are by women, seven by men; the gendered nature of the 'cry for help' is still inconsistent. Stengel, Cook and Kreeger, *Attempted Suicide:* 58
144. Stengel, Cook and Kreeger, *Attempted Suicide:* 82.
145. Ibid. 86, emphasis in original.
146. Ibid. 22, emphasis in original.

3   Self-Harm Becomes Epidemic: Mental Health (1959) and Suicide (1961) Acts

1. E. Stengel, 'Enquiries into Attempted Suicide (Abridged)' *Proceedings of the Royal Society of Medicine* 45 (1952): 620
2. R. Jack, *Women and Attempted Suicide* London, Lawrence Erlbaum (1992): xii
3. Hansard HC Deb 31 October 1958, Vol 594, col 521
4. H. Matthew, 'Poisoning in the Home by Medicaments' *British Medical Journal* 2, 5517 (1966): 788
5. C. Millard and S. Wessely, 'Parity of Esteem Between Physical and Mental Health' *British Medical Journal* 349, 7894 (2014): 6821; P. Border and Chris Millard, Parliamentary Office of Science and Technology 'Parity of Esteem in Mental Health' (2015) online at: http://www.parliament.uk/business/publications/research/briefing-papers/POST-PN-485/parity-of-esteem-for-mental-health accessed 30 January 2015
6. K. Robinson, 'The Public and Mental Health' in *Trends in the Mental Health Services* H.L. Freeman and W.A.J. Farndale (eds) Oxford, Pergamon (1963): 16

7. A.T. Scull, *Decarceration: Community Treatment and the Deviant: A Radical View* London, Prentice-Hall (1977)

8. H. Lester and J. Glasby, *Mental Health Policy and Practice* Basingstoke, Palgrave Macmillan (2006): 27

9. This starts early: R.M. Titmuss, 'Community Care: Fact or Fiction?' in *Commitment to Welfare* R.M. Titmuss (ed.) London: George Allen & Unwin, ([1961] 1968): 221–5; see also H.R. Rollin, 'Social and Legal Repercussions of the Mental Health Act, 1959' *British Medical Journal* 1, 5333 (1963): 788.

10. A. Rogers and D. Pilgrim, *Mental Health Policy in Britain* 2nd ed. London, Macmillan (2001): 55, 65. Recent examples of 'institution-community' binaries include M. Gorsky, 'The British National Health Service 1948–2008: A Review of the Historiography' *Social History of Medicine* 21(3) (2008): 449; Lester and Glasby, *Mental Health Policy and Practice:* 27

11. N. Rose, quoted in Rogers and Pilgrim, *Mental Health Policy in Britain:* 73

12. G. Eghigian, 'Deinstitutionalizing the History of Contemporary Psychiatry' *History of Psychiatry* 22(2) (2011): 203, emphasis in the original.

13. For an apt historiographical summary, see J. Welshman, 'Rhetoric and Reality: Community Care in England and Wales, 1948–74' in *Outside the Walls of the Asylum: The History of Care in the Community 1750–2000* P. Bartlett and D. Wright (eds) London: Athlone Press (1999): 205. This covers the contributions of Kathleen Jones, Andrew Scull, Peter Sedgwick and Joan Busfield.

14. C. Webster, 'Psychiatry and the Early National Health Service: The Role of the Mental Health Standing Advisory Committee' in *150 Years of British Psychiatry, 1841–1991* H. Freeman and G. E. Berrios (eds) London, Gaskell (1991): 104

15. Baron Percy of Newcastle, 'The Report of the Royal Commission of the Law Relating to Mental Illness and Mental Deficiency' London, HMSO (1957)

16. 'Mental Health Act, 1959' London: HMSO (1959): 3

17. B. Wootton, *Social Science and Social Pathology* London, George Allen & Unwin (1959): 208–9

18. D. Stafford-Clark, 'Attempted Suicide' *Lancet* 281, 7278 (1963): 448–9. Two of the other participants in this correspondence argument are Neil Kessel and Richard Asher, and the correspondence was initially sparked by an Erwin Stengel article.

19. N. Kessel, 'Self-Poisoning – Part II'*British Medical Journal* 2, 5474 (1965): 1340

20. G.D. Middleton, D.W. Ashby and F. Clark, 'An Analysis of Attempted Suicide in an Urban Industrial District' *Practitioner* 187 (1961): 776–82; M. Woodside, 'Attempted Suicides Arriving at a General Hospital' *British Medical Journal* 2, 5093 (1958): 411–4; J.J. Fleminger and B.L. Mallett, 'Psychiatric Referrals from Medical and Surgical Wards' *British Journal of Psychiatry* 108 (1962): 183–90

21. J.E. Lennard-Jones and R.A.J. Asher, 'Why Do They Do It? A Study of Pseudocide' *Lancet* 1, 7083 (1959): 1138

22. Lennard-Jones and Asher, 'Study of Pseudocide': 1138–9

23. Ibid. 1138

24. M. Jarvis, *Conservative Governments, Morality and Social Change in Affluent Britain, 1957–64* Manchester, Manchester University Press (2005): 6

25. Baron Wolfenden of Westcott, *Report of the Committee on Homosexual Offences and Prostitution* London, HMSO (1957)
26. P. Hennessy, *Having It So Good : Britain in the Fifties* London, Allen Lane (2006): 505–6
27. Jarvis, *Conservative Governments:* 11
28. H.L.A. Hart, quoted in Hennessy, *Having It So Good:* 510
29. Macmillan to Butler, 24 June 1961 Discussions on Act to amend law relating to suicide TNA: PREM 11/3241; Jarvis, *Conservative Governments:* 95; Hennessy, *Having It So Good:* 510.
30. Draft Reply Macmillan to Butler, 'Suicide and Attempted Suicide: Proposals to Amend Law' Suicide and attempted suicide: proposals to amend law, 1958–1961 TNA: HO 291/141; Jarvis, *Conservative Governments:* 95; Hennessy, *Having It So Good:* 510. However facetious Macmillan might appear to Butler, he is not alone in failing to appreciate the link between the law on suicide and the treatment of attempted suicide. (Lord) Lewis Silkin admits that he is unaware 'from the Bill itself that it is intended to cover also attempted suicide.' HL Deb 02 March 1961 Vol 229 cols 253–4
31. Jarvis, *Conservative Governments:* 95–6
32. Hennessy, *Having It So Good:* 510
33. Jarvis, *Conservative Governments:* 98
34. Hennessy, *Having It So Good:* 510
35. HC Deb 06 February 1958 Vol 581 col 1327
36. British Medical Association Committee on Psychiatry and the Law, 'The Law Relating to Attempted Suicide' *British Medical Journal Supplement* 2210, 5406 (1947): S.103. One of the arguments advanced by a Church of England booklet (see below) is similar: 'The punishment of the offender is not likely to deter others from attempting to commit suicide, if only because they will be confident of success.' Church Information Office, *Ought Suicide to Be a Crime? A Discussion on Suicide, Attempted Suicide and the Law* Westminster, Church Information Office (1959): 10. Note the use of 'would-be suicide' in a context that implies earnest, uncomplicated intent.
37. G.L. Williams, *The Sanctity of Life and the Criminal Law: On Contraception, Sterilization, Artificial Insemination, Abortion, Suicide and Euthanasia* London, Faber & Faber (1958): 250, 253; quoting W.L. Neustatter, *Psychological Disorder and Crime* London, Christopher Johnson (1953): 68
38. Williams, *Sanctity of Life:* 255
39. *Ought Suicide?:* 12, 21–2
40. HL Deb 01 July 1959 Vol 217 col 599
41. HC Deb 06 March 1958 Vol 583 col 1342; HC Deb 13 March 1958 Vol 584 col 74W
42. HC Deb 22 May 1958 Vol 588 col 1471; HC Deb 31 October 1958 Vol 594 col 523. Again, note the association of 'would-be suicide' with genuine intent.
43. HM(61)94 'Attempted Suicide' (18 September 1961) in Circulars letters notes and memoranda (1961) TNA: MH 119/15
44. Notes of Meeting at Home Office with H.O., Ministry of Health, Magistrates' Association and BMA, dated 13 March 1959 TNA: MH 137/383
45. Anon., 'Obituary: Professor Linford Rees' *Daily Telegraph* (11 September 2004)
46. J. Fry, *Casualty Services and Their Setting: A Study in Medical Care* Nuffield Provincial Hospitals Trust (1960); Ministry of Health, 'Accident and

Emergency Services: Report of a Sub-Committee (Platt Report)' London, HMSO (1962)

47. Although patients were 'in many instances, referred to a psychiatrist later', 'the availability of psychiatric help in assessing the case is relatively infrequent.' W.L. Rees and J.S. Stead 'Report to Council' Law in relation to attempted suicide: general correspondence, 1958–61 TNA: MH 137/383

48. Joint Committee of the British Medical Association and the Magistrates' Association, 'The Law and Practice in Relation to Attempted Suicide in England and Wales' London, British Medical Association and Magistrates' Association (1958) 9 in Suicide and attempted suicide: proposals to amend law, 1958–1961 TNA: HO 291/141

49. Note for File for Maclay by Benner dated 28 November 1960 TNA: MH137/383

50. Ibid. It is invoked again, two months later: Note for Emery by Benner dated 13 February 1961 TNA: MH 137/383

51. Note for Dr. Macdonald from T.E. Nodder dated 3 January 1961, TNA: PRO: MH137/383

52. Note from Emery to 'Deputy Secretary [of State for Health]' dated 15 February 1961 TNA: MH 137/383

53. E. Stengel, 'Attempted Suicide: Its Management in the General Hospital' *Lancet* 1, 7275 (1963): 233

54. HM(61)94 'Attempted Suicide' TNA: MH 119/15

55. Stengel, 'Attempted Suicide: Management': 233

56. Note to Mr. Dodds, Dr. Goodman and Deputy Secretary. From Benner[?] dated 3 August 1961 TNA: MH 137/383

57. E. Stengel 'The National Health Service and the Suicide Problem' in *Sociological Review Monograph No. 5: Sociology and Medicine: Studies within the Framework of the British National Health Service* P. Halmos (ed.) Keele: University of Keele (1962): 205

58. Letter to 'Secretary' from Hardwick, dated 28 December 1961 TNA: MH137/384

59. Stengel, 'Attempted Suicide: Management': 235

60. Note from P. Benner to G.C. Tooth dated 14 August 1962 TNA: MH 137/384. Tooth is most famous for the 1961 *Lancet* paper co-authored with Eileen Brooke which claimed that the demand for inpatient beds in British mental hospitals would halve – acting as a springboard for Enoch Powell's *Hospital Plan* (1962) (see Chapter 5); G.C. Tooth and E.M. Brooke, 'Trends in the Mental Hospital Population and Their Effect on Future Planning'*Lancet* 1, 7179 (1961)

61. Note by Benner for Mr. Emery dated 7 September 1962; Note from G.C. Tooth to P. Benner noted 15 August 1962 TNA: PRO: MH 137/384

62. Circular letter 'Attempted Suicide' (Ref 94600/1/49b) dated 28 September 1962 TNA: MH 137/384

63. 'Attempted Suicide' in 'Attempted Suicide – Hospital Treatment and Replies to Questionnaire' TNA: MH 150/220

64. 'Attempted Suicide' TNA: MH 150/220

65. 'Summary of Replies to questionnaire [Sheffield RHB] as the result of Ministry of Health letter 94600/1/49B dated 28th September 1962' enclosed with letter from Sheffield RHB to Ministry of Health dated 19 December 1962 TNA: MH

150/220. This form is possibly the one mentioned by Parkin and Stengel in 1965 (see Chapter 5).

66. 'Summary of Replies to Questionnaire [Sheffield RHB]' TNA: MH 150/220
67. 'North West Metropolitan Hospital Board – Treatment of Attempted Suicide' TNA: MH 150/220
68. Letter from North West Metropolitan Hospital Board to Ministry of Health dated 31 December 1962 TNA: MH 150/220
69. Letter from Wessex RHB to Ministry of Health dated 2 January 1963 TNA: MH 150/220
70. 'Attempted Suicide [Welsh Hospital Board]' enclosed with letter from Welsh Hospital Board to Ministry of Health dated 21 January 1963 TNA: MH 220/150
71. Letter from Stengel to Dunbar dated 14 September 1962 TNA: MH 150/220. 'Diagnostic index' presumably refers to the one used for the Hospital In-Patient Enquiry (HIPE).
72. Letter from Stengel to Dunbar dated 14 September 1962 TNA: MH 150/220
73. W.M. Millar, G. Innes and G.A. Sharp, 'Hospital and Outpatient Clinics: The Design of a Reporting System and the Difficulties to Be Expected in the Execution' in The Burden on the Community: The Epidemiology of Mental Illness A Symposium (1962): 2
74. P. Sainsbury and J. Grad, 'Evaluation of Treatment and Services' in The Burden on the Community: The Epidemiology of Mental Illness: A Symposium (1962) appendix I, unnumbered page
75. Note for Tooth from Otley dated 18 December 1964 in 'Attempted Suicide – Research Proposed by Department' and Note from Brothwood to Perry dated 17 May 1965 TNA: MH 150/221
76. I. Hacking, *Rewriting the Soul* Princeton, Princeton University Press (1995): 236

4 Self-Harm as a Result of Domestic Distress

1. For example P. Sedgwick, *Psychopolitics* London, Pluto Press (1982): 104; A. Rogers and D. Pilgrim, *Mental Health Policy in Britain* 2nd ed. London, Macmillan (2001): 64–5
2. E. Powell, *Emerging Patterns for the Mental Health Services and the Public* London, National Association for Mental Health (1961)
3. J.D.N. Hill, 'Review of Policy on Psychiatric Research – Summary prepared by Dr. Denis Hill of the proposals made in his talk to Council on the 16 January 1959'; 16 January 1959: review of psychiatric research by Dr J D N Hull [*sic*], 1959 TNA: FD 9/91; see also G.M. Carstairs, *This Island Now* Harmondsworth, Penguin (1963) and H.R. Rollin, 'Carstairs, George Morrisson (1916–1991)' in *Oxford Dictionary of National Biography* London, Oxford University Press (2004)
4. N. Kessel, 'Psychiatric Morbidity in a London General Practice' *British Journal of Preventive & Social Medicine* 14 (1960): 16–22
5. N. Kessel and M. Shepherd, 'Neurosis in Hospital and General Practice' *Journal of Mental Science* 108 (1962): 159–66; N. Kessel and H.J. Walton, *Alcoholism* Harmondsworth, Penguin (1965); N. Kessel and G. Grossman, 'Suicide in Alcoholics' *British Medical Journal* 2, 5268 (1961): 1671–2

6. B. Deakin, G. Hay, D. Goldberg and B. Hore, 'William Ivor Neil Kessel' *Psychiatric Bulletin* 28 (2004): 309
7. N. Kessel, 'Self-Poisoning – Part I' *British Medical Journal* 2, 5473 (1965): 1265
8. For example, I.R.C. Batchelor and M.B. Napier, 'Broken Homes and Attempted Suicide' *British Journal of Delinquency* 4 (1953): 101
9. N. Kessel and W. McCulloch, 'Repeated Acts of Self-Poisoning and Self-Injury' *Proceedings of the Royal Society of Medicine* 59(2) (1966): 89 referring to N. Kessel, W. McCulloch, J. Hendry, D. Leslie, I. Wallace and R. Webster, 'Hospital Management of Attempted Suicide in Edinburgh' *Scottish Medical Journal* 9 (1964)
10. Kessel, 'Self-Poisoning – Part I': 1265
11. N. Kessel and E.M. Lee, 'Attempted Suicide in Edinburgh' *Scottish Medical Journal* 7 (1962): 130
12. H.J.S. Matthew and A.A.H. Lawson, *Treatment of Common Acute Poisonings* Edinburgh, E. & S. Livingstone (1967): 2
13. For example, Anon., 'Attempted Suicide: Changes in English Law Wanted' *British Medical Journal* 1, 5081 (1958)
14. In fact, Scotland has its own Mental Health (Scotland) Act (1960), but these are treated as almost identical by some researchers. See for example M. Woodside, 'Are Observation Wards Obsolete? A Review of One Year's Experience in an Acute Male Psychiatric Admission Unit' *British Journal of Psychiatry* 114 (1968): 1013. In any case, the point here is about *publicity* and *visibility*, something much more relevant to the 1959 Act.
15. A.B. Sclare and C.M. Hamilton, 'Attempted Suicide in Glasgow' *British Journal of Psychiatry* 109 (1963): 609, 614
16. N. Kessel, W. McCulloch and E. Simpson, 'Psychiatric Service in a Centre for the Treatment of Poisoning' *British Medical Journal* 2, 5363 (1963): 987
17. Kessel, McCulloch and Simpson, 'Psychiatric Service': 987
18. Kessel et al., 'Hospital Management': 334; McCulloch and Philip, argue later that 'persons unskilled in psychiatry tend to equate the severity of the physical state with the severity of the underlying problem. There is no such easy equation'. W. McCulloch and A.E. Philip, *Suicidal Behaviour* Oxford, Pergamon (1972): 31
19. N. Kessel, 'Self-Poisoning – Part II' *British Medical Journal* 2, 5474 (1965): 1339
20. N. Kessel, 'The Respectability of Self-Poisoning and the Fashion of Survival' *Journal of Psychosomatic Research* 10(1) (1966): 35
21. Kessel, 'Self-Poisoning – Part II': 1338
22. Kessel et al., 'Hospital Management': 333; Kessel, 'Self-Poisoning – Part II': 1339
23. Kessel, 'Respectability': 30
24. In the early 1960s, he works under 'an operational definition that included all cases of overdosage, gassing or injury admitted to the ward, where it could be established that these were self-inflicted.' Kessel and Lee, 'Attempted Suicide in Edinburgh': 130
25. A. Proudfoot and L.F. Lescott, 'Henry Matthew: The Father of Modern Clinical Toxicology' *Journal of the Royal College of Physicians of Edinburgh* 39 (2009): 358

26. S. Locket, 'Barbiturate Deaths' *British Medical Journal* 2, 5162 (1959): 1332
27. E. Stengel, N.G. Cook and I.S. Kreeger, *Attempted Suicide: Its Social Significance and Effects* London, Chapman and Hall (1958): 113
28. Kessel, 'Respectability': 30
29. Standing Medical Advisory Committee of the Central Health Services Council, 'Report of the Sub Committee: Emergency Treatment in Hospital of Cases of Acute Poisoning' [Atkins Report] London, HMSO (1962). There is also a report by the Scottish Bureau.
30. Dr. Dooley 'The Problem of Accidental Poisoning in England and Wales and Arrangements for Prevention and Treatment' [Atkins] Committee papers, 1959 TNA: MH 133/230
31. Standing Medical Advisory Committee of the Central Health Services Council, 'Hospital Treatment of Acute Poisoning. Report of the Joint Sub-Committee of the Standing Medical Advisory Committees' [Hill Report] London, HMSO (1968)
32. H. Matthew, A.T. Proudfoot, S.S. Brown and R.C. Aitken, 'Acute Poisoning: Organization and Work-Load of a Treatment Centre' *British Medical Journal* 3, 5669 (1969): 489–92
33. N. Kreitman, 'The Coal Gas Story. United Kingdom Suicide Rates, 1960–71' *British Journal of Preventive & Social Medicine* 30(2) (1976): 86–93; C. Hassall and W.H. Trethowan, 'Suicide in Birmingham' *British Medical Journal* 1, 5802 (1972): 717–8
34. Stengel, Cook and Kreeger, *Attempted Suicide:* 111
35. Kessel, 'Self-Poisoning – Part I': 1269
36. C.A.H. Watts, *Depressive Disorders in the Community* Bristol, John Wright & Sons (1966): 131. Attempted suicide ceases to be a *common law misdemeanour* when suicide ceases to be a felony in 1961.
37. K. Dunnell and A. Cartwright, *Medicine Takers, Prescribers and Hoarders* London: Routledge & Kegan Paul (1972)
38. For example, C.M. Callahan and G.E. Berrios, *Reinventing Depression: A History of the Treatment of Depression in Primary Care, 1940–2004* Oxford, Oxford University Press (2005)
39. J. Botting, 'The History of Thalidomide' *Drug News & Perspectives* 15(9) (2002): 604–11
40. W.R. Brain, 'Drug Addiction: Report of the Interdepartmental Committee' London, Ministry of Health (1961); W.R. Brain, 'Drug Addiction: The Second Report of the Interdepartmental Committee' London, Ministry of Health (1965)
41. Drug Dependence Advisory Committee, 'Cannabis: Report (Wootton Report)' London, HMSO (1968); HC Deb 27 January 1969 Vol 776 col 959. This phrase is widely misquoted as 'rising tide of permissiveness'
42. Stengel, Cook and Kreeger, *Attempted Suicide* 116; Woodside, 'Attempted Suicides': 411
43. Kessel, 'Self-Poisoning – Part II': 1336
44. McCulloch and Philip, *Suicidal Behaviour:* 31, emphasis in the original
45. N. Kessel, 'Neurosis & the N.H.S.' *The Twentieth Century* CLXXII, 1015 (1962): 55
46. R.A.J. Asher, 'Fashions in Disease' *The Twentieth Century* CLXXII, 1015 (1962): 18

47. E.S. Paykel, B.A. Prusoff and J.K. Myers, 'Suicide Attempts and Recent Life Events. A Controlled Comparison' *Archives of General Psychiatry* 32(3) (1975): 327–33

48. Compare: Stengel, Cook and Kreeger, *Attempted Suicide*: 117 to E. Stengel, *Suicide and Attempted Suicide* Harmondsworth, Penguin Books (1964): 11–12. For an historical discussion of this topic see E. Ramsden and D. Wilson, 'The Nature of Suicide: Science and the Self-Destructive Animal' *Endeavour* 34(1) (2010): 21–4

49. D. Armstrong, *Political Anatomy of the Body: Medical Knowledge in Britain in the Twentieth Century* Cambridge, Cambridge University Press (1983): 106

50. R. Hayward, 'Sadness in Camberwell: Imagining Stress and Constructing History in Post War Britain [Draft]', paper given in Washington D.C., 2010; G.W. Brown and T. Harris, *Social Origins of Depression : A Study of Psychiatric Disorder in Women* London, Tavistock Publications (1978)

51. Jack, *Women and Attempted Suicide*: 21

52. J. Faulkner, 'Note on visit to the Unit for Research on the Epidemiology of Psychiatric Illness Edinburgh' 9 November 1962 dated 29 January 1963: 1–2 Visits by HQ staff: progress report for 1964, 1962–1974 TNA: FD 12/408

53. See: E.L. Younghusband, *Social Work in Britain: A Supplementary Report on the Employment and Training of Social Workers* Dunfermline, Carnegie United Kingdom Trust (1951): 81.

54. J.P. Nursten, 'Editor's Foreword' to A. Munro and J.W. McCulloch, *Psychiatry for Social Workers* 2nd ed. Oxford, Pergamon Press (1975): xiii

55. Kessel and Lee, 'Attempted Suicide in Edinburgh': 130 and note

56. J.K. Wing, 'Survey Methods and the Psychiatrist' in *Methods of Psychiatric Research* P. Sainsbury and N. Kreitman (eds) London, Oxford University Press (1963): 118

57. Kessel, McCulloch and Simpson, 'Psychiatric Service': 985, 987

58. N. Timms, *Psychiatric Social Work in Great Britain, 1939–1962* London, Routledge & Kegan Paul (1964): 117

59. A. Munro and J.W. McCulloch, *Psychiatry for Social Workers* 2nd ed. Oxford, Pergamon Press (1975): 68

60. Munro and McCulloch, *Psychiatry for Social Workers*: 73, 76, 201–2

61. N. Kessel, 'Attempted Suicide' *Medical World* 97 (1962): 314

62. J. Busfield, *Men, Women and Madness: Understanding Gender and Mental Disorder* Basingstoke, Macmillan (1996): 190

63. N. Kessel and W. McCulloch, 'Repeated Acts of Self-Poisoning and Self-Injury' *Proceedings of the Royal Society of Medicine* 59(2) (1966): 91

64. Kessel, 'Self-Poisoning – Part I': 1268

65. Ibid. 1337

66. For a general history, see J. Lewis, 'Public Institution and Private Relationship: Marriage and Marriage Guidance, 1920–1968' *Twentieth Century British History* 1(3) (1990): 233–63

67. E. Irvine, 'Psychiatric Social Work' in E. Younghusband, *Social Work in Britain, 1950–1975: A Follow-Up Study* London, Allen & Unwin (1978): 194

68. N. Kreitman, 'Mental Disorder in Married Couples' *British Journal of Psychiatry* 108 (1962): 438–46; N. Kreitman, 'The Patient's Spouse' *British Journal of Psychiatry* 110 (1964): 159–73; L.S. Penrose, 'Mental Illness in Husband and Wife: A Contribution to the Study of Assortive Mating in Man' *Psychiatric*

Quarterly 18 (1944): 161–6; E.T.O. Slater and M. Woodside, *Patterns of Marriage: A Study of Marriage Relationships in the Urban Working Classes* London, Cassell (1951): 12–14
69. Kessel and Lee, 'Attempted Suicide in Edinburgh': 134
70. Kessel, 'Self-Poisoning – Part II': 1337
71. Timms, *Psychiatric Social Work:* 117
72. E. Heimler, *Mental Illness and Social Work* Harmondsworth, Penguin (1967): 118–19, emphasis in the original.
73. Munro and McCulloch, *Psychiatry for Social Workers*: 75–6
74. Kessel, 'Self-Poisoning – Part I': 1269
75. Ibid.
76. McCulloch and Philip, *Suicidal Behaviour:* 20.
77. J.H. Wallis, *Marriage Guidance: A New Introduction* London, Routledge & Kegan Paul (1968): 105
78. See also, M. Shepherd, 'Morbid Jealousy: Some Clinical and Social Aspects of a Psychiatric Symptom' *British Journal of Psychiatry* 107 (1961): 687–753
79. Jack, *Women and Attempted Suicide:* ix
80. Ali Haggett, 'Housewives, Neuroses and the Domestic Environment in Britain, 1945–1970' in *Health and the Modern Home* M. Jackson (ed.) Abingdon, Routledge (2007): 84; see also R. Cooperstock and H.L. Lennard, 'Some Social Meanings of Tranquilizer Use' *Sociology of Health and Illness* 1(3) (1979): 331–47
81. Kessel, Self-Poisoning – Part I': 1267
82. Slater and Woodside, *Patterns of Marriage:* 15
83. A classic statement is E. Showalter, *The Female Malady: Women, Madness and English Culture 1830–1980* London, Virago (1987)
84. Kessel, 'Self-Poisoning – Part II': 1336–7
85. Kessel and Lee, 'Attempted Suicide in Edinburgh': 134
86. N. Kessel, 'Attempted Suicide' *Lancet* 281, 7278 (1963): 448
87. Kessel, 'Self-Poisoning – Part I': 1267
88. Ibid.
89. M. Shepherd, A.C. Brown, B. Cooper and G. Kalton, *Psychiatric Illness in General Practice* London, Oxford University Press (1966): 149
90. N. Rose, *Governing the Soul : The Shaping of the Private Self* London, Free Association Press (1999): 180
91. J. Stewart, '"I Thought You Would Want to Come and See His Home": Child Guidance and Psychiatric Social Work in Inter-War Britain' in *Health and the Modern Home* M. Jackson (ed.) London, Routledge (2007): 117
92. N. Timms, *Psychiatric Social Work in Great Britain, 1939–1962* London, Routledge & Kegan Paul (1964): 126
93. J.B.S. Lewis quoted in Timms, *Psychiatric Social Work:* 112–13
94. Timms, *Psychiatric Social Work:* 113
95. J. Pickstone, 'Psychiatry in District General Hospitals' in *Medical Innovations in Historical Perspective* J. Pickstone (ed.) (1992): 186; M. Gorsky, 'The British National Health Service 1948–2008: A Review of the Historiography' *Social History of Medicine* 21(3) (2008): 450
96. W.S. Maclay, 'The Adolf Meyer Lecture: A Mental Health Service' *The American Journal of Psychiatry* 120 (1963): 215
97. C.P. Seager, *Psychiatry for Nurses Social Workers and Occupational Therapists* London, Heinemann Medical (1968): 212

98. Ministry of Health, 'A Hospital Plan for England and Wales' London, HMSO (1962) quoted in J.C. Little, 'A Rational Plan for Integration of Psychiatric Services to an Urban Community' *Lancet* 2, 7318 (1963): 1159

99. Little, 'Rational Plan': 1159

100. P.K. Bridges, K.M. Koller, and T.K. Wheeler, 'Psychiatric Referrals in a General Hospital' *Acta Psychiatrica Scandinavica* 42(2) (1966): 171

101. See, among very many examples: C.P.B. Brook and D. Stafford-Clark, 'Psychiatric Treatment in General Wards' *Lancet* 277, 7187 (1961): 1159; J. Bierer, 'Day Hospitals and Community Care' *Comprehensive Psychiatry* 4(6) (1963): 381

102. Maclay, 'A Mental Health Service': 211

103. H.L. Freeman, 'Psychiatry in the National Health Service 1948–1998' *British Journal of Psychiatry* 175 (1999): 3

104. E.W. Dunkley and E. Lewis, 'North Wing: A Psychiatric Unit in a General Hospital' *Lancet* 281, 7273 (1963): 156; D.R. Benady and J. Denham, 'Development of an Early Treatment Unit from an Observation Ward' *British Medical Journal* 2, 5372 (1963): 1569

105. D.K. Henderson, *The Evolution of Psychiatry in Scotland* Edinburgh, E. & S. Livingstone (1964): 116

106. R.P. Snaith and S. Jacobson, 'The Observation Ward and the Psychiatric Emergency' *British Journal of Psychiatry* 111 (1965): 25

107. For example Dunkley and Lewis, 'North Wing': 156

108. J. Carson and E.H. Kitching, 'Psychiatric Beds in a General Ward: A Year's Experience' *Lancet* 1, 6559 (1949): 833, 835

109. E. Stengel, 'Attempted Suicide: Its Management in the General Hospital' *Lancet* 1, 7275 (1963): 234

110. J.G. Macleod and H.J. Walton, 'Liaison between Physicians and Psychiatrists in a Teaching Hospital' *Lancet* 294, 7624 (1969): 790–1

111. R.W. Crocket, 'In-Patient Care of General Hospital Psychiatric Patients' *British Medical Journal* 2, 4828 (1953): 123

112. Anon., 'Psychiatry in the General Hospital' *Lancet* 279, 7239 (1962): 1107

113. Maclay, 'Appraisal of English Psychiatry': 106

114. Pickstone, 'District General Hospitals': 190

115. For details of the Brighton study, see: Snaith and Jacobson, 'Observation Ward'; for Bristol, see J. Roberts and D. Hooper, 'The Natural History of Attempted Suicide in Bristol' *British Journal of Medical Psychology* 42(4) (1969): 303–12; for Leicester, see G.G. Ellis, K.A. Comish and R.L. Hewer, 'Attempted Suicide in Leicester' *Practitioner* 196 (1966): 557–61

116. D. Parkin and E. Stengel, 'Incidence of Suicidal Attempts in an Urban Community' *British Medical Journal* 2, 5454 (1965): 133–4

117. Parkin and Stengel, 'Urban Community': 133–4

118. Kessel, McCulloch, and Simpson, 'Psychiatric Service': 987

119. Parkin and Stengel, 'Urban Community': 133

120. Ibid. emphasis in the original.

121. Parkin and Stengel, 'Urban Community': 137

122. GPs are involved with the Stengel et al. 1958 study, in the Dulwich General Hospital group (Group V), where '[w]ith 33 patients ... a medical practitioner had been involved in the admission. With 26 ... no general practitioner had intervened, and with 17 patients ... it was uncertain whether this was so'.

The Sheffield study is different; practitioners are being utilised to provide the object without the necessity for a general hospital. Stengel, Cook and Kreeger, *Attempted Suicide:* 96
123. Watts, *Depressive Disorders:* 127
124. E. Jones, S. Rahman and R. Woolven, 'The Maudsley Hospital: Design and Strategic Direction, 1923–1939' *Medical History* 51(3) (2007): 358, 369; Mayou, 'General Hospital Psychiatry': 768
125. Bridges, Koller and Wheeler, 'Psychiatric Referrals': 178, 180
126. P.K. Bridges and K.M. Koller, 'Attempted Suicide. A Comparative Study' *Comprehensive Psychiatry* 7(4) (1966): 240
127. P.K. Bridges, 'Psychiatric Emergencies' *Postgraduate Medical Journal* 43, 503 (1967): 599
128. N. Kreitman, A.E. Philip, S. Greer and C.R. Bagley, 'Parasuicide' *British Journal of Psychiatry* 115 (1969): 746–7
129. S. Greer, J.C. Gunn and K.M. Koller, 'Aetiological Factors in Attempted Suicide' British Medical Journal 2, 5526 (1966): 1352
130. Greer, Gunn and Koller, 'Aetiological Factors': 1352, 1354–5
131. S. Greer and J.C. Gunn, 'Attempted Suicides from Intact and Broken Parental Homes' *British Medical Journal* 2, 5526 (1966): 1355–7
132. I.S. Kreeger, 'Initial Assessment of Suicidal Risk' *Proceedings of the Royal Society of Medicine* 59(2) (1966): 93–4
133. Kreeger, 'Initial Assessment': 92, 94, 96.
134. Ellis, Comish and Hewer, 'Attempted Suicide in Leicester': 560.
135. Bridges, Koller and Wheeler, 'Psychiatric Referrals': 178–80
136. Kessel and McCulloch, 'Repeated Acts': 92, emphasis in the original. Stengel curiously undercuts the case for conflict over beds, arguing that '[m]any ['attempted suicides'] will be given an appointment in the psychiatric outpatient clinic…hardly more than one out of five, require to be transferred to a psychiatric department for further treatment.' Stengel, 'Attempted Suicide: Management': 235
137. As Krohn expresses it: '"split and inverted" in the suggestive phrase of Latour and Woolgar'; R. Krohn, 'Why Are Graphs So Central in Science?' *Biology and Philosophy* 6(2) (1991): 197
138. I. Hacking, *Rewriting the Soul* Princeton, Princeton University Press (1995): 236
139. Kreitman et al., 'Parasuicide': 146
140. See: Reconstitution of Unit: memo to MRC on research in progress and copy of the refounding proposal from Dr Kreitman, 1970–1972 TNA: FD 12/412

5   Self-Harm as Self-Cutting: Inpatients and Internal Tension

1. A.A.H. Lawson and I. Mitchell, 'Patients with Acute Poisoning Seen in a General Medical Unit (1960–71)' *British Medical Journal* 4, 5833 (1972): 153
2. A.J. Smith, 'Self-Poisoning with Drugs: a Worsening Situation' *British Medical Journal* 4, 5833 (1972): 157–9

3. T.A. Holding, D. Buglass, J.C. Duffy and N. Kreitman, 'Parasuicide in Edinburgh – A Seven-Year Review 1968–1974' *British Journal of Psychiatry* 130 (1977): 534; K. Hawton, J. O'Grady, M. Osborn and D. Cole, 'Adolescents who Take Overdoses: Their Characteristics, Problems and Contacts with Helping Agencies' *British Journal of Psychiatry* 140 (1982): 118–23

4. S. Chaney, 'Self-Mutilation and Psychiatry: Impulse, Identity and the Unconscious in British Explanations of Self-Inflicted Injury, c. 1864–1914' PhD Thesis, University College London (2013)

5. I.R.C. Batchelor, 'Attempted Suicide' *British Medical Journal* (1955): 595

6. N. Kessel, 'Attempted Suicide' *Medical World* 97 (1962): 312

7. H.G. Morgan, Helen Pocock and Susan Pottle, 'The Urban Distribution of Non-Fatal Deliberate Self-Harm' *British Journal of Psychiatry* 126 (1975): 320

8. N. Kreitman (ed.), *Parasuicide* London, John Wiley & Sons (1977): 8

9. See C. Millard, 'Self-Mutilation: Emergence, Exclusions and Contexts 1967–1976' MA Thesis, University of York (2007); C. Millard, 'Making the Cut: The Production of "Self-harm" in Anglo-Saxon Psychiatry' *History of the Human Sciences* 26(2) (2013): 126–50; B.J. Brickman, '"Delicate" Cutters: Gendered Self-mutilation and Attractive Flesh in Medical Discourse' *Body and Society* 10(4) (2004): 87–111

10. M.A. Simpson, 'Self Mutilation and Suicide' in *Suicidology: Contemporary Developments* E.S. Shneidman (ed.) New York, Grune and Stratton (1976): 310

11. The most cited articles are: H. Graff and R. Mallin, 'The Syndrome of the Wrist Cutter' *American Journal of Psychiatry* 124 (1967): 36–42; L. Crabtree Jr., 'A Psychotherapeutic Encounter with a Self-mutilating Patient' *Psychiatry* 30 (1967): 91–100; H. Grunebaum and G. Klerman, 'Wrist Slashing' *American Journal of Psychiatry* 124 (1967): 527–34; P. Pao, 'The Syndrome of Delicate Self-cutting' *British Journal of Medical Psychology* 42 (1969): 195–205; J.S. Kafka, 'The Body as Transitional Object: A Psychoanalytic Study of a Self-mutilating Patient' *British Journal of Medical Psychology* 42 (1969): 207–12; S. Asch, 'Wrist Scratching as a Symptom of Anhedonia: A Predepressive State' *Psychoanalytic Quarterly* 40 (1971): 603–13; R. Rosenthal, C. Rinzler, R. Wallsch, and E. Klausner, 'Wrist-cutting syndrome: the meaning of a gesture' *American Journal of Psychiatry* 128 (1972): 1363–8

12. American Psychiatric Association, 'Non-Suicidal Self-Injury' *Diagnostic and Statistical Manual of Mental Disorders* (5th ed.) Washington, D.C., American Psychiatric Association (2013): 803

13. D. Tantam and N. Huband, *Understanding Repeated Self-Injury: A Multidisciplinary Approach* Basingstoke, Palgrave Macmillan (2009): 1

14. L. Fagin, 'Repeated Self-Injury: Perspective from General Psychiatry' *Advances in Psychiatric Treatment* 12 (2006): 193

15. B. Walsh and P. Rosen, *Self-mutilation: Theory, Research and Treatment* Guildford, Guildford Press (1988): 32, quoted in A. Favazza, *Bodies Under Siege: Self-Mutilation, Nonsuicidal Self-Injury, and Body Modification in Culture and Psychiatry* 3rd ed. Baltimore, Johns Hopkins University Press (2011): 198–9

16. J. Sutton, *Healing the Hurt Within: Understand Self-Injury and Self-Harm, and Heal the Emotional Wounds* 3rd ed. Oxford, How To Books (2007): 14

17. N. Pengelly, B. Ford, P. Blenkiron and S. Reilly, 'Harm Minimisation after Repeated Self-harm: Development of a Trust Handbook' *Psychiatric Bulletin* 32 (2008): 63

18. Favazza, *Bodies Under Siege* 3rd ed.: 198
19. Royal College of Psychiatrists, 'Self-harm, Suicide and Risk: Helping People Who Self-harm' College Report CR158 (2010) online at: http://www.rcpsych. ac.uk/files/pdfversion/cr158.pdf accessed 30 January 2015: 6
20. K. Hawton, K.E.A. Saunders and R.C. O'Connor, 'Self-harm and Suicide in Adolescents' *Lancet* 379, 9834 (2012): 2373–4
21. Sutton, *Healing the Hurt Within:* 105–6
22. P.A. Adler and P. Adler, *The Tender Cut: Inside the Hidden World of Self-Injury* New York, New York University Press (2011); T. McShane *Blades, Blood and Bandages* Basingstoke, Palgrave Macmillan (2012); P. McCormick, *Cut* London, Collins Flamingo (2009); C. Rainfield, *Scars* Lodi, New Jersey, Westside Books (2011)
23. Shelly A. James, 'Has Cutting Become Cool? Normalising, Social Influence and Socially Motivated Deliberate Self-Harm in Adolescent Girls' Doctor of Clinical Psychology Research Project, Massey University, Albany, New Zealand (2013)
24. S. Chaney, '"A Hideous Torture on Himself": Madness and Self-Mutilation in Victorian Literature' *Journal of Medical Humanities* 32 (2011): 280. Some twentieth-century clinical studies deal with self-mutilation by people labelled as 'subnormal' or 'defective' (what would now be called learning difficulties) characterised by repetitive head-banging or self-biting. There is also significant literature on self-mutilation in people with Lesch-Nyhan or Cornelia De Lange syndromes, which are serious and severely inhibiting chromosomal and genetic conditions. This is significantly different, with only minimal relevance to the emergence of the contemporary event of self-mutilation stereotypes.
25. A. Strauss, L. Schatzman, R. Bucher, D. Ehrlich and M. Sabshin, *Psychiatric Ideologies and Institutions* London, Collier-Macmillan Limited (1964): 12
26. D. Offer and P. Barglow, 'Adolescent and Young Adult Self-Mutilation Incidents in a General Psychiatric Hospital' *Archives of General Psychiatry* 3 (1960): 194
27. Offer and Barglow, 'Self-Mutilation Incidents' (1960): 195
28. See A.H. Stanton and M.S. Schwartz, *The Mental Hospital* London, Tavistock Publications (1954)
29. Offer and Barglow, 'Self-mutilation Incidents' (1960): 201
30. M.E. Staub, *Madness Is Civilization: When the Diagnosis Was Social, 1948–1980* London, University of Chicago Press (2011)
31. T.F. Main, 'The Ailment' *Medical Psychology* 30(3) (1957): 129–45
32. W. Dorrell 'Dr Colin McEvedy' *Independent* 30 August 2005 online at: http://www.independent.co.uk/news/obituaries/dr-colin-mcevedy-8715136.html accessed 30 January 2015; C. Richmond, 'Colin McEvedy' *British Medical Journal* 331 (2005): 847
33. P.D. Moss and C.P. McEvedy, 'An Epidemic of Overbreathing Among Schoolgirls' *British Medical Journal* 2 (1966): 1295–1300; C.P. McEvedy, A. Griffith and T. Hall, 'Two School Epidemics' *British Medical Journal* 2 (1966): 1300–2
34. C.P. McEvedy, 'Self-Inflicted Injuries' Diploma of Psychological Medicine Dissertation, University of London (1963): 25
35. McEvedy, 'Self-inflicted Injuries': 7
36. Ibid. 28

37. McEvedy, 'Self-inflicted Injuries' p. 8. There is much historical work to be done on how slashed wrists become associated with suicide. Sarah Chaney informs me from her studies of self-mutilation in Victorian asylums that cut throats are judged as suicidal, but not cuts anywhere else (with the extremely occasional exception concerning veins on the inside of the elbow). McEvedy himself refers to 'popular opinion, perhaps fed by vague memories of the suicidal techniques of the ancient roman aristocracy, continues to hold the belief that an injury [to the wrists] will prove rapidly fatal'. It would be very interesting to chart when the idea of wrist-cutting became suggestive of suicide – gestural or otherwise. Unfortunately, the scope of the present chapter precludes such a study.

38. McEvedy, 'Self-inflicted Injuries': 21

39. Ibid. 12

40. Ibid. 9, 14, 23

41. Ibid. 28–9

42. Ibid. 13, 28, 8

43. Ibid. 29

44. For example, Royal College of Psychiatrists 'Helping People Who Self-Harm': 6, 21

45. C.P. McEvedy and A.W. Beard, 'Royal Free Epidemic of 1955: A Reconsideration' *British Medical Journal* 1, 5687 (1970): 7–11

46. For example, K. Skegg, 'Self-harm' *Lancet* 366, 9495 (2005): 1471–83

47. An article on this subject from the late 1960s argues that 'we feel that this [prison mutilation] is less a reflection of a personality pattern of turning anger inward than it is a reaction to confinement'. J.L. Claghorn and D.R. Beto 'Self-Mutilation in a Prison Mental Hospital' *Journal of Social Therapy* 13 (1967): 140

48. D.W. McKerracher, D.R.K. Street and L.J. Segal, 'A Comparison of the Behaviour Problems Presented by Male and Female Subnormal Offenders' *British Journal of Psychiatry* 112 (1966): 891–7; D.W. McKerracher, T. Loughnane and R.A. Watson 'Self-Mutilation in Female Psychopaths' *British Journal of Psychiatry* 114 (1968): 829–32

49. F. Graham 'Probability of Detection and Institutional Vandalism' *British Journal of Criminology* 21(4) (1981): 361

50. McKerracher, Street and Segal, 'Comparison of Behaviour Problems': 896

51. Ibid.

52. The terms 'psychopath' and 'psychopathy' are broad and loosely defined both in the 1960s and today. Martyn Pickersgill has written of the 'profound uncertainties and ambivalences [that] are evident even within expert discourse' on psychopathy. M. Pickersgill, 'Psyche, Soma, and Science Studies: New Directions in the Sociology of Mental Health and Illness' *Journal of Mental Health* 19(4)(2010): 388; see also: M. Pickersgill, 'Between Soma and Society: Neuroscience and the Ontology of Psychopathy' *BioSocieties* 4 (2009): 45–60

53. McKerracher, Loughnane and Watson 'Self-mutilation': 829

54. Ibid. 830

55. Ibid. 831

56. Ibid. 830–1

57. P.C. Matthews 'Epidemic Self-Injury in an Adolescent Unit' *International Journal of Social Psychiatry* 14 (1968): 131

58. G.A. Kelly, *The Psychology of Personal Constructs* 2 Vols, New York, Norton (1955)
59. J. P. Watson, 'Relationship between a Self-mutilating Patient and her Doctor' *Psychotherapy and Psychosomatics* 18 (1970): 67–8
60. Watson 'Self-mutilating Patient': 70
61. Ibid. 71
62. See Brickman, '"Delicate" Cutters' and Millard, 'Making the Cut'
63. A.M.R. Keller 'Self-Mutilation' MSc Thesis, University of London, Institute of Psychiatry (1970) 4
64. Keller, 'Self-Mutilation':2, 8, 12–13
65. B.R. Ballinger, 'Minor Self-Injury' *British Journal of Psychiatry* 118 (1971): 535
66. Ballinger, 'Minor Self-Injury': 536–7
67. For example, A.R. Gardner and A.J. Gardner, 'Self-mutilation, Obsessionality and Narcissism' *British Journal of Psychiatry* 127 (1975): 127–32; Alec Roy, 'Self-Mutilation' *British Journal of Medical Psychology* 51 (1978): 201–3
68. D. Bhugra, *Culture and Self Harm: Attempted Suicide in South Asians in London* London, Psychology Press (1994)
69. S.S.A. Waldenberg, 'Wrist-Cutting: A Psychiatric Enquiry' MPhil Thesis, University of London, Institute of Psychiatry (1972): 12.
70. Waldenberg, 'Wrist-Cutting': 38
71. See Millard, 'Making the Cut': 137–40
72. Waldenberg 'Wrist-Cutting' Abstract (unnumbered page)
73. Waldenberg, 'Wrist-Cutting': 3. He does mention the 'much less profuse British literature on the subject', citing only McEvedy, 'Self-inflicted Injuries', and McKerracher, Loughnane and Watson, 'Self-mutilation'. Waldenberg, 'Wrist-Cutting': 4–5
74. Waldenberg, 'Wrist-Cutting': 24
75. Pao, quoted in Waldenberg, 'Wrist-Cutting': 2
76. Waldenberg, 'Wrist-Cutting': 3, 5
77. Waldenberg, 'Wrist-Cutting': 1. Here it seems that Waldenberg has accepted that the behaviour is attention-seeking, but it becomes clear later that this is not a point pressed in the dissertation.
78. Waldenberg, 'Wrist-Cutting': 38, 42–3
79. Ibid. 1, 30
80. Ibid. 37, 43
81. Ibid. 40, 43–4
82. Ibid. 43–4
83. Ibid. 13, 39
84. S. Crown and A.H. Crisp, 'A Short Clinical Diagnostic Self-rating Scale for Psychoneurotic Patients: The Middlesex Hospital Questionnaire (M.H.Q.)' *British Journal of Psychiatry* 112 (1966): 917–23; J. Sandler, 'Studies in Psychopathology using at Self-Assessment Inventory. I. The Development and Construction of the Inventory' *British Journal of Medical Psychology* 27 (1954): 147–52
85. Gardner and Gardner, 'Self-mutilation': 127
86. Ibid. 128–9
87. Ibid. 131

88. M.A. Simpson, 'Phenomenology of Self-Mutilation in a General Hospital Setting' *Canadian Psychiatric Association Journal* 20(6) (1975): 429–34; M.A. Simpson, 'Self-Mutilation and Suicide' in *Suicidology Contemporary Developments* E.S. Schneidman (ed.) (1976): 286–315; M.A. Simpson, *Medical Education: A Critical Approach* London, Butterworths (1972); M.A. Simpson, 'Self-Mutilation and Borderline Syndrome' *Dynamische Psychiatrie* 1 (1977): 42–8

89. Simpson 'Phenomenology': 429–30

90. Ibid. 430, 432

91. Ibid. 431–2

92. Ibid. 290, 295

93. Ibid. 296–7, 310

94. Alec Roy, 'Self-mutilation' *British Journal of Medical Psychology* 51 (1978): 201, 203.

95. For example, K. Hawton, K. Rodham, E. Evans and R. Weatherall, 'Deliberate Self Harm in Adolescents: Self Report Survey in Schools in England' *British Medical Journal* 325 (2002): 1207–11; K. Hawton and M. Goldacre, 'Hospital Admissions for Adverse Effects of Medicinal Agents (Mainly Self-Poisoning) Among Adolescents in the Oxford Region' *British Journal of Psychiatry* 141 (1982): 166–70

96. D. Pallis, A. Langley and D. Birtchnell, 'Excessive Use of Psychiatric Services by Suicidal Patients' *British Medical Journal* 3, 5977 (1975): 216

97. K. Hawton, 'Excessive Use of Psychiatric Services' *British Medical Journal* 3, 5983 (1975): 595

98. K. Hawton, 'Deliberate Self-poisoning and Self-injury in the Psychiatric Hospital' *British Journal of Medical Psychology* 51(3) (1978): 257–8

99. Hawton, 'Deliberate Self-poisoning and Self-injury': 258

100. Sutton, *Healing the Hurt Within:* 112

101. M.D. Enoch and W.H. Trethowan, *Uncommon Psychiatric Syndromes* 2nd ed. Bristol, John Wright and Sons (1979): 87–8

102. H.G. Morgan, *Death Wishes? The Understanding and Management of Deliberate Self-Harm* Chichester, John Wiley and Sons (1979): 115–16

103. Morgan, *Death Wishes?* 116–17

104. Ibid. 117

105. This includes five groups (plus an 'other' category): cuts to the wrists or forearms, other cuts, gunshot/drowning/asphyxiation (including hanging), jumping from a height, and jumping in front of a moving vehicle. K. Hawton and J. Catalán, *Attempted Suicide* 1st ed. Oxford, Oxford University Press (1982): 117

106. Hawton and Catalán, *Attempted Suicide* 1st ed.: 116

107. Ibid. 118–19, 125

108. Hawton and Catalán *Attempted Suicide* 2nd ed. Oxford, Oxford University Press (1987): 16

109. H.G. Morgan, C.J. Burns-Cox, H. Pocok and Susan Pottle, 'Deliberate Self-Harm: Clinical and Socio-Economic Characteristics of 368 Patients' *British Journal of Psychiatry* 127 (1975): 572

110. N. Kreitman (ed.), *Parasuicide* London, John Wiley & Sons (1977): 8

111. R. Turner and H. Morgan, 'Patterns of Health Care in Non-fatal Deliberate Self-harm' *Psychological Medicine* 9(3) (1979): 488, referring to Morgan et al., 'Deliberate self-harm' (1975)

112. Hawton, 'Deliberate Self-poisoning and Self-injury': 257

113. Morgan, *Death Wishes?* 116
114. Waldenberg, 'Wrist-Cutting': 38
115. Gardner and Gardner 'Self-Mutilation': 129
116. Waldenberg, 'Wrist-Cutting': 2, 64–5
117. Watson 'Self-mutilating Patient': 67
118. For a recent example, see I. Ciorba, O. Farcus, R. Giger and L. Nisa, 'Facial Self-mutilation: An Analysis of Published Cases' *Postgraduate Medical Journal* 90, 1062 (2014): 191–200
119. Waldenberg, 'Wrist-Cutting': 20.
120. McEvedy,'Self-inflicted Injuries': 8
121. Simpson, 'Phenomenology': 430
122. Rinzler and Shapiro, 'Wrist-Cutting and Suicide': 487; Nelson and Grunebaum, 'A Follow-up Study': 1348
123. Hawton and Catalán, *Attempted Suicide* 2nd ed.: 152

Conclusion: The Politics of Self-Harm: Social Setting and Self-Regulation

1. H. Kushner, 'Biochemistry, Suicide and History: Possibilities and Problems' *Journal of Interdisciplinary History* 16(1) (1985): 71
2. Roger Cooter, 'Neural Veils and the Will to Historical Critique' *Isis* 105(1) (2014): 145
3. E.T.O. Slater, 'Psychiatry in the 'Thirties' *Contemporary Review* 226, 1309 (1975): 75
4. C. Millard, 'Reinventing Intention: "Self-Harm" and the "Cry for Help" in Post-War Britain' *Current Opinion in Psychiatry* 25(6) (2012): 503–7
5. B. Walsh and P. Rosen, *Self-Mutilation: Theory, Research and Treatment* Guildford, Guildford Press (1988); two more recent statements are: J. Sutton, *Healing the Hurt Within: Understand Self-Injury and Self-Harm, and Heal the Emotional Wounds*, 3rd ed. Oxford, How To Books (2007); and K. Skegg, 'Self-harm' *Lancet* 366, 9495 (2005): 1471–83
6. Skegg, 'Self-harm': 1473
7. R. Gardner, R. Hanka, V.C. O'Brien, A.J. Page and R. Rees, 'Psychological and social evaluation in cases of deliberate self-poisoning admitted to a general hospital' *British Medical Journal* 2, 6102 (1977): 1567–70; R. Gardner, R. Hanka, B. Evison, P.M. Mountford, V.C. O'Brien and S.J. Roberts, 'Consultation-liaison scheme for self-poisoned patients in a general hospital' *British Medical Journal* 2, 6149 (1978): 1392–4; J. Catalan, P. Marsack, K.E. Hawton, D. Whitwell, J. Fagg and J.H.J. Bancroft, 'Comparison of doctors and nurses in the assessment of self-poisoning patients' *Psychological Medicine* 10 (1980): 483–91; J.G.B Newson-Smith, 'Use of social workers as alernatives to psychiatrists in assessing parasuicide' in *The Suicide Syndrome* R. Farmer and S. Hirsch (eds) London, Croom Helm (1980): 215–25; D.R Blake and M.G. Bramble, 'Self-poisoning: psychiatric assessment by junior staff' *British Medical Journal* 1, 6180 (1979): 1763
8. Department of Health and Social Security, 'Health Services Management – The Management of Deliberate Self-Harm' Health Notice HN(84)25 (1984) Health circulars and notices, local authority circulars and social service letters, medical letters and other guidance material TNA: BN 1/118

9. R. Porter, 'The Patient's View: Doing Medical History from Below' *Theory and Society* 14(2) (1985): 175–98
10. J.W. Scott, 'The Evidence of Experience' *Critical Enquiry* 17(4) (1991): 777
11. For the former, see A. Bechara, 'The neurology of social cognition' *Brain: A Journal of Neurology* 125(8) (2002): 1673–5; for the latter, see M. Pickersgill 'Between soma and society: Neuroscience and the ontology of psychopathy' *BioSocieties* 4 (2009): 45–60
12. H. Rose and S. Rose, *Genes, Cells and Brains: The Promethean Promises of the New Biology* London, Verso (2012): 22–3
13. M.A. Simpson, 'Self Mutilation and Suicide' in *Suicidology: Contemporary Developments* E.S. Shneidman (ed) New York, Grune and Stratton (1976) 309–10
14. F. Gardner, *Self-Harm: A Psychotherapeutic Approach* Hove, Brunner-Routledge (2001) 26
15. S.R. Schroeder, M.L. Oster-Granite and T. Thompson, *Self-Injurious Behavior: Gene-Brain-Behavior-Relationships* American Psychological Association, Washington, D.C. (2002)
16. D. Brent, 'Prevention of self harm in adolescents' *British Medical Journal* 342 (2011): 592
17. B. Stanley and L. Siever, 'The interpersonal dimension of borderline personality disorder: toward a neuropeptide model' *American Journal of Psychiatry* 167 (2010): 24–39
18. M. Anguelova, C. Benkelfat and G. Turecki, 'A systematic review of association studies investigating genes coding for serotonin receptors and the serotonin transporter. II: Suicidal behavior' *Molecular Psychiatry* 8 (2003): 646–53; J.G. Keilp, M.A. Oquendo, B.H. Stanley, A.K. Burke, T.B. Cooper, K.M. Malone and J.J. Mann, 'Future suicide attempt and responses to serotonergic challenge.' *Neuropsychopharmacology* 35 (2010): 1063–72
19. A.R. Favazza, *Bodies Under Siege: Self-Mutilation, Nonsuicidal Self-Injury, and Body Modification in Culture and Psychiatry* 3rd ed. Baltimore, Johns Hopkins University Press (2011) x
20. P.A. Adler and P. Adler, 'The De-Medicalization of Self-Injury: From Psychopathology to Sociological Deviance' *Journal of Contemporary Ethnography* 36 (2007): 537–70
21. W. Bareiss, '"Mauled by a Bear": Narrative analysis of self-injury among adolescents in US news, 2007–2012' *Health: An Interdisciplinary Journal for the Social Study of Health, Illness and Medicine* 18(3) (2014): 279–301
22. Cooter, 'Neuro Veils': 153
23. D. Keay, interviewing Margaret Thatcher, 'Aids, Education and the Year 2000!' *Woman's Own* (31 October 1987) Transcript online at http://www.margaretthatcher.org/document/106689 accessed 28 January 2015
24. A. McSmith, *No Such Thing as Society* London, Constable (2010) 4
25. McSmith, *No Such Thing as Society* (2010) 5
26. M. Foucault, *The Birth of Biopolitics: Lectures at the College de France, 1978–1979* A.I. Davidson (ed.) and Graham Burchell (trans.) Basingstoke, Palgrave Macmillan (2008) 78–9
27. Foucault, *Birth of Biopolitics* 116
28. P. Anderson, *Spectrum: From Left to Right in the World of Ideas* London, Verso (2005) 54

29. I. Butler and M. Drakeford *Social Policy, Social Welfare and Scandal: How British Public Policy is Made* Basingstoke, Palgrave Macmillan (2003) 200
30. Rose and Rose, *Genes, Cells and Brains* 20
31. R. Duschinsky and L.A. Rocha, 'Introduction: The problem of the family in Foucault's work' in *Foucault the Family and Politics* R. Duschinsky and L.A. Rocha (eds) Basingstoke, Palgrave Macillan (2012) 12
32. G. Harkins, 'Foucault, the family and the cold monster of neoliberalism' in *Foucault the Family and Politics* R. Duschinsky and L.A. Rocha (eds) Basingstoke, Palgrave Macillan (2012) 87
33. An example used by Jan Sutton is: L. McKinstry, 'This "Epidemic" Is All Selfishness' *Daily Telegraph* (1 August 2004)
34. K.S. Rajan, *Biocapital: The Constitution of Postgenomic Life* Durham, NC, Duke University Press (2006) 3
35. Rose and Rose, *Genes, Cells and Brains* 2
36. Rajan, *Biocapital* 3
37. L. Daston and P. Gallison, *Objectivity* London, Zone Books (2005)
38. S. Shapin and S. Schaffer, *Leviathan and the Air-Pump: Hobbes, Boyle and the Experimental Life* Princeton, Princeton University Press (1985); B. Latour and S. Woolgar, *Laboratory Life: The Social Construction of Scientific Facts* London, Sage Publications (1979)
39. D. Smail, *On Deep History and the Brain* Berkeley: University of California Press (2008). For some critical context, see M. Littlefield and J. Johnson (eds), *The Neuroscientific Turn: Transdisciplinarity in the Age of the Brain* Ann Arbor, University of Michigan Press (2012)
40. G. Eaton, accessed 'Balls reaffirms Labour's commitment to cuts in 2015' *The Staggers: New Statesman's Rolling Politics Blog* online at: http://www.newstatesman.com/politics/2013/12/balls-reaffirms-labours-commitment-cuts-2015 accessed 30 January 2015
41. S. Bell, *When Salem Came to the Boro: the True Story of the Cleveland Child Abuse Crisis* London, Pan (1988)

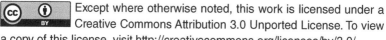 Except where otherwise noted, this work is licensed under a Creative Commons Attribution 3.0 Unported License. To view a copy of this license, visit http://creativecommons.org/licenses/by/3.0/

Bibliography

Unpublished Works

Theses

S. Chaney, 'Suicide, Mental Illness and the Asylum: The Case of Bethlem Royal Hospital 1845–1875' MA Dissertation, University College London (2009)

S. Chaney, 'Self-Mutilation and Psychiatry: Impulse, Identity and the Unconscious in British Explanations of Self-Inflicted Injury, c. 1864–1914' PhD Thesis, University College London (2013)

S.A. James, 'Has Cutting Become Cool? Normalising, Social Influence and Socially-Motivated Deliberate Self-Harm in Adolescent Girls' Doctor of Clinical Psychology Research Project, Massey University, Albany, New Zealand (2013)

A.M.R. Keller, 'Self-Mutilation' MSc Thesis, University of London, Institute of Psychiatry (1970)

C.P. McEvedy, 'Self-Inflicted Injuries' Diploma of Psychological Medicine Dissertation, University of London, Institute of Psychiatry (1963)

C. Millard, 'Self-Mutilation: Emergence, Exclusions and Contexts 1967–1976' MA Thesis, University of York (2007)

S.S.A. Waldenberg, 'Wrist-Cutting: A Psychiatric Enquiry' MPhil Thesis, University of London, Institute of Psychiatry (1972)

National Archives, Kew

BN 1/118: Health circulars and notices, local authority circulars and social service letters, medical letters and other guidance material

FD 7/1043: Proposed Unit for Research into the Epidemiology of Mental Diseases, University College Hospital, London, 1958–9

FD 9/91: 16th January 1959: review of psychiatric research by Dr J D N Hull [sic], 1959

FD 12/407: Future of the Unit: transfer of work to Edinburgh, 1960–7

FD 12/408: Visits by HQ staff: progress report for 1964, 1962–74

FD 12/409: Progress report for 1964: report to Council of work carried out up to 1964, 1964–5

FD 12/412: Reconstitution of Unit: memo to MRC on research in progress and copy of the refounding proposal from Dr Kreitman, 1970–2

HO 45/24439: CRIMINAL; POLICE: Removal of attempted suicides to hospital: police liability to guard patient to prevent another attempt, 1907–50

HO 291/141: Suicide and attempted suicide: proposals to amend law, 1958–61

MEPO 2/6955: Suicides and attempted suicides: numbers recorded annually and analysis of motives and methods used, 1933–1954. MEPO 2/10121: The Suicide Act 1961: police action in cases of attempted suicide, 1961–71

MEPO 3/2436: Attempted suicide: police supervision of offenders in hospital, 1916–51

MH 95/32: Maudsley Hospital, Denmark Hill, London SE5, 1923–48

MH 119/15: Circulars, letters, notes and memoranda (indexed), 1961

MH 133/230: [Atkins] Committee papers, 1959

MH 137/383: Law in relation to attempted suicide: general correspondence, 1958–61

MH 137/384: Law in relation to attempted suicide: proposed amendments, 1961–2

MH 150/220: Attempted suicide: hospital board statistics on numbers of patients treated, 1962–3

MH 150/221: Attempted suicide: proposed research, 1964–5

PREM 11/3241: Discussions on Act to amend law relating to suicide

Lothian Health Services Archive, University of Edinburgh, Edinburgh

LHB 1/59/7: Proposed Regional Poisoning Treatment Centre; revaluation of functions of Ward 3; poisoning among children, 1962

Conference Papers

R. Hayward, 'Sadness in Camberwell: Imagining Stress and Constructing History in Post War Britain [Draft]' paper given in Washington, D.C., 2010

Published Works

A.D., 'Obituary Notices: Hon. W.S. Maclay' *British Medical Journal* 1, 5392 (1964)

P.A. Adler and P. Adler, 'The De-Medicalization of Self-Injury: From Psychopathology to Sociological Deviance' *Journal of Contemporary Ethnography* 36 (2007): 537–70

P.A. Adler and P. Adler, *The Tender Cut: Inside the Hidden World of Self-Injury* New York, New York University Press (2011)

D. Aldridge, *Suicide: The Tragedy of Hopelessness* London, Jessica Kingsley (1998)

American Psychiatric Association, 'Non-Suicidal Self-Injury' *Diagnostic and Statistical Manual of Mental Disorders* 5th ed. Washington, D.C., American Psychiatric Association (2013)

O. Anderson, *Suicide in Victorian and Edwardian England* Oxford, Clarendon (1987)

O. Anderson, 'Prevention of Suicide and Parasuicide: What Can We Learn from History?' *Journal of the Royal Society of Medicine* 82 (11) (1989): 640–2

P. Anderson, *Spectrum: From Left to Right in the World of Ideas* London, Verso (2005)

M. Anguelova, C. Benkelfat and G. Turecki, 'A Systematic Review of Association Studies Investigating Genes Coding for Serotonin Receptors and the Serotonin Transporter. II: Suicidal Behavior' *Molecular Psychiatry* 8 (2003): 646–53

Anon., 'Attempted Suicide: Changes in English Law Wanted' *British Medical Journal* 1, 5081 (1958)

Anon., 'Psychiatry in the General Hospital' *Lancet* 279, 7239 (1962): 1107

Anon., 'Obituary: Professor Linford Rees' *Daily Telegraph* (11 September 2004)

Anon., 'Lost Hospitals of London' online at: http://ezitis.myzen.co.uk/stfrancis.html accessed 21 August 2012

D. Armstrong, *Political Anatomy of the Body: Medical Knowledge in Britain in the Twentieth Century* Cambridge, Cambridge University Press (1983)

S.S. Asch, 'Wrist Scratching as a Symptom of Anhedonia: A Predepressive State' *Psychoanalytic Quarterly* 40 (1971): 603–13.

R.A.J. Asher, 'Arrangements for the Mentally Ill' *The Lancet* 268, 6955 (1956): 1265–6

R.A.J. Asher, 'Fashions in Disease' *The Twentieth Century* CLXXII, 1015 (1962): 18–19

P.G. Aungle, 'Sir Ivor Batchelor' *Psychiatric Bulletin* 29 (2005): 439

B.R. Ballinger, 'Minor Self-Injury' *British Journal of Psychiatry* 118 (1971): 535–8

W. Bareiss, '"Mauled by a Bear": Narrative Analysis of Self-injury among Adolescents in US News, 2007–2012' *Health: An Interdisciplinary Journal for the Social Study of Health, Illness and Medicine* 18(3) (2014): 279–301

P. Barham, *Closing the Asylum: The Mental Patient in Modern Society* Harmondsworth, Penguin (1992)

I.R.C. Batchelor, 'Psychopathic States and Attempted Suicide' *British Medical Journal* 1, 4875 (1954): 1342–7

I.R.C. Batchelor, 'Repeated Suicidal Attempts' *British Journal of Medical Psychology* 27(3) (1954): 158–63

I.R.C. Batchelor, 'Alcoholism and Attempted Suicide' *Journal of Mental Science* 100, 419 (1954): 451–61

I.R.C. Batchelor, 'Attempted Suicide' *British Medical Journal* 1, 4913 (1955): 595–7

I.R.C. Batchelor, 'Management and Prognosis of Suicidal Attempts in Old Age' *Geriatrics* 10(6) (1955): 291–3

I.R.C. Batchelor, *Henderson and Gillespie's Textbook of Psychiatry for Students and Practitioners* 10th ed. London, Oxford University Press (1969)

I.R.C. Batchelor and M.B. Napier, 'Broken Homes and Attempted Suicide' *British Journal of Delinquency* 4 (1953): 99–108

I.R.C. Batchelor and M.B. Napier, 'The Sequelae and Short-Term Prognosis of Attempted Suicide; the Results of a One-Year Follow-up of 200 Cases' *Journal of Neurology, Neurosurgery & Psychiatry* 17(4) (1954): 261–6

I.R.C. Batchelor and M.B. Napier, 'Attempted Suicide in Old Age' *British Medical Journal* 2, 4847 (1953): 1186–90

A. Bechara, 'The Neurology of Social Cognition' *Brain: A Journal of Neurology* 125(8) (2002): 1673–5

S. Bell, *When Salem Came to the Boro: The True Story of the Cleveland Child Abuse Crisis* London, Pan (1988)

D.R. Benady and J. Denham, 'Development of an Early Treatment Unit from an Observation Ward' *British Medical Journal* 2, 5372 (1963): 1569–72

G.E. Berrios, *The History of Mental Symptoms: Descriptive Psychopathology since the Nineteenth Century* Cambridge, Cambridge University Press (1996)

D. Bhugra, *Culture and Self Harm: Attempted Suicide in South Asians in London* London, Psychology Press (1994)

J. Bierer, 'Day Hospitals and Community Care' *Comprehensive Psychiatry* 4(6) (1963): 381–6

W.R. Bion, *Experiences in Groups* London, Tavistock Publications (1961)

D.R. Blake and M.G. Bramble, 'Self-poisoning: Psychiatric Assessment by Junior Staff' *British Medical Journal* 1, 6180 (1979): 1763

M. Borch-Jacobsen, 'Making Psychiatric History: Madness as Folie À Plusieurs' *History of the Human Sciences* 14(2) (2001): 19–38

P. Border and C. Millard, Parliamentary Office of Science and Technology 'Parity of Esteem in Mental Health' (2015) online at: http://www.parliament.uk/business/publications/research/briefing-papers/POST-PN-485/parity-of-esteem-for-mental-health accessed 30 January 2015

J. Botting, 'The History of Thalidomide' *Drug News & Perspectives* 15(9) (2002): 604–11

J. Bourke, *Dismembering the Male: Men's Bodies, Britain and the Great War* London, Reaktion Books (1996)

J. Bowlby, *Maternal Care and Mental Health* Geneva, World Health Organization (1951)

W.R. Brain, 'Drug Addiction: Report of the Interdepartmental Committee' London, Ministry of Health (1961)

W.R. Brain, 'Drug Addiction: The Second Report of the Interdepartmental Committee' London, Ministry of Health (1965)

D. Brent, 'Prevention of Self Harm in Adolescents' *British Medical Journal* 342 (2011): 592–3

B.J. Brickman, '"Delicate" Cutters: Gendered Self-mutilation and Attractive Flesh in Medical Discourse' *Body and Society* 10(4) (2004): 87–111

P.K. Bridges, 'Psychiatric Emergencies' *Postgraduate Medical Journal* 43, 503 (1967): 599–604

P.K. Bridges and K.M. Koller, 'Attempted Suicide. A Comparative Study' *Comprehensive Psychiatry* 7(4) (1966): 240–7

P.K. Bridges, K.M. Koller, and T.K. Wheeler, 'Psychiatric Referrals in a General Hospital' *Acta Psychiatrica Scandinavica* 42(2) (1966): 171–82

British Medical Association Committee on Psychiatry and the Law, 'The Law Relating to Attempted Suicide' *British Medical Journal Supplement* 2210, 5406 (1947): S.103

C.P.B. Brook and D. Stafford-Clark, 'Psychiatric Treatment in General Wards' *Lancet* 277, 7187 (1961): 1159–62

J. Busfield, *Men, Women and Madness: Understanding Gender and Mental Disorder* Basingstoke, Macmillan (1996)

E.N. Butler, 'Observation Units' *Proceedings of the Royal Society of Medicine* 33 (1940): 725–32

I. Butler and M. Drakeford, *Social Policy, Social Welfare and Scandal: How British Public Policy is Made* Basingstoke, Palgrave Macmillan (2003)

G.W. Brown, E.M. Monck, G.M. Carstairs and J.K. Wing, 'Influence of Family Life on the Course of Schizophrenic Illness' *British Journal of Preventive and Social Medicine* 16(2) (1962): 55–68

G.W. Brown and T. Harris, *Social Origins of Depression: A Study of Psychiatric Disorder in Women* London, Tavistock Publications (1978)

P. Brown, *The Transfer of Care: Psychiatric Deinstitutionalization and Its Aftermath* Boston: Routledge & Kegan Paul (1985)

C.M. Callahan and G.E. Berrios, *Reinventing Depression: A History of the Treatment of Depression in Primary Care, 1940–2004* Oxford, Oxford University Press (2005)

D. Campbell, 'Overcrowded Hospitals "Killed 500" Last Year, Claims Top A&E Doctor' *Observer* (24 January 2015) online at: http://www.theguardian.com/uk-news/2015/jan/24/overcrowded-hospitals-deaths accessed 26 January 2015

S.S. Canetto, 'Men Who Survive a Suicidal Act: Successful Coping or Failed Masculinity?' in *Men's Health and Illness: Gender, Power and the Body*, D. Sabo and D.F. Gordon (eds) Thousand Oaks, Sage (1995): 292–304

W.B. Cannon, *Bodily Changes in Pain, Hunger, Fear and Rage* New York, D. Appleton & Co. (1929)

J. Carson and E.H. Kitching, 'Psychiatric Beds in a General Ward: A Year's Experience' *Lancet* 1, 6559 (1949): 833–5

G.M. Carstairs, *This Island Now* Harmondsworth, Penguin (1963)

J. Catalan, P. Marsack, K.E. Hawton, D. Whitwell, J. Fagg and J.H.J. Bancroft, 'Comparison of Doctors and Nurses in the Assessment of Self-poisoning Patients' *Psychological Medicine* 10 (1980): 483–91

E.F. Catford, *The Royal Infirmary of Edinburgh 1929–1979* Edinburgh, Scottish Academic Press (1984)

S. Chaney, '"A Hideous Torture on Himself": Madness and Self-Mutilation in Victorian Literature' *Journal of Medical Humanities* 32 (2011): 279–89

Church Information Office, *Ought Suicide to Be a Crime? A Discussion on Suicide, Attempted Suicide and the Law* Westminster, Church Information Office (1959)

I. Ciorba, O. Farcus, R. Giger and L. Nisa, 'Facial Self-mutilation: An Analysis of Published Cases' *Postgraduate Medical Journal* 90, 1062 (2014): 191–200

J.L. Claghorn and D.R. Beto, 'Self-Mutilation in a Prison Mental Hospital' *Journal of Social Therapy* 13 (1967): 133–41

R. Cooperstock and H.L. Lennard, 'Some Social Meanings of Tranquilizer Use' *Sociology of Health and Illness* 1(3) (1979): 331–47

R. Cooter, 'The End? History-Writing in the Age of Biomedicine (and Before)' in R. Cooter with C. Stein *Writing History in the Age of Biomedicine* New Haven and London, Yale University Press (2013): 1–40

R. Cooter, 'Neural Veils and the Will to Historical Critique' *Isis* 105(1) (2014): 145–54

L.H. Crabtree Jr., 'A Psychotherapeutic Encounter with a Self-mutilating Patient' *Psychiatry* 30 (1967): 91–100

R.W. Crocket, 'In-Patient Care of General Hospital Psychiatric Patients' *British Medical Journal* 2, 4828 (1953): 122–5

S. Crown, 'Post-War Maudsley Personalities' *Psychiatric Bulletin* 12(7) (1988): 263–6

S. Crown and A.H. Crisp, 'A Short Clinical Diagnostic Self-rating Scale for Psychoneurotic Patients: The Middlesex Hospital Questionnaire (M.H.Q.)' *British Journal of Psychiatry* 112 (1966): 917–23

L. Daston and P. Gallison, *Objectivity* London, Zone Books (2005)

B. Deakin, G. Hay, D. Goldberg and B. Hore, 'William Ivor Neil Kessel' *Psychiatric Bulletin* 28 (2004): 309–10

Department of Health, 'Your Very Good Health' [video] (1948) online at: http://www.nhs.uk/Livewell/NHS60/Pages/VideointroducingthenewNHS.aspx accessed 11 July 2012

Department of Health, 'The Configuration of Mental Health Services' (2007) online at: http://www.dh.gov.uk/en/Publicationsandstatistics/Publications/PublicationsPolicyAndGuidance/Browsable/DH_4897913 accessed 13 July 2012

W. Dorrell, 'Dr Colin McEvedy', *Independent* 30 August 2005 online at: http://www.independent.co.uk/news/obituaries/dr-colin-mcevedy-8715136.html accessed 30 January 2015

Drug Dependence Advisory Committee, 'Cannabis: Report (Wootton Report)' London, HMSO (1968)

D.M. Dunlop, 'Foreword' to H.J.S. Matthew and A.A.H. Lawson, *Treatment of Common Acute Poisonings* Edinburgh, E. & S. Livingstone (1967)

E.W. Dunkley and E. Lewis, 'North Wing: A Psychiatric Unit in a General Hospital' *The Lancet* 281, 7273 (1963): 156–9

K. Dunnell and A. Cartwright, *Medicine Takers, Prescribers and Hoarders* London: Routledge and Kegan Paul (1972)

R. Duschinsky and L.A. Rocha, 'Introduction: The Problem of the Family in Foucault's Work' in *Foucault, the Family and Politics* R. Duschinsky and L.A. Rocha (eds) Basingstoke, Palgrave Macmillan (2012): 1–38

G.G. Ellis, K.A. Comish and R.L. Hewer, 'Attempted Suicide in Leicester' *Practitioner* 196 (1966): 557–61

F. Graham, 'Probability of Detection and Institutional Vandalism' *British Journal of Criminology* 21(4) (1981): 361–5

D.F. Early, 'The Changing Use of the Observation Ward' *Public Health* 76(5) (1962): 261–8

G. Eaton, 'Balls Reaffirms Labour's Commitment to Cuts in 2015' *The Staggers: New Statesman's Rolling Politics Blog* online at: http://www.newstatesman.com/politics/2013/12/balls-reaffirms-labours-commitment-cuts-2015

G. Eghigian, 'Deinstitutionalizing the History of Contemporary Psychiatry' *History of Psychiatry* 22(2) (2011): 201–14

G. Eghigian, A. Killen, and C. Leuenberger, 'The Self as Project: Politics and the Human Sciences in the Twentieth Century' *Osiris* 22(1) (2007): 1–25

M.D. Eilenberg, M.J. Pritchard and P.B. Whatmore, 'A 12-Month Survey of Observation Ward Practice' *British Journal of Preventive and Social Medicine* 16 (1962): 22–9

M.D. Eilenberg and P.B. Whatmore, 'Police Admissions to a Mental Observation Ward' *Medicine, Science, and the Law* 2 (1961): 96–100

M.D. Enoch and W.H. Trethowan, *Uncommon Psychiatric Syndromes* 2nd ed. Bristol, John Wright and Sons (1979)

L. Fagin, 'Repeated Self-Injury: Perspective from General Psychiatry' *Advances in Psychiatric Treatment* 12 (2006): 193–201

L. Faithfull, 'Younghusband, Dame Eileen Louise (1902–1981)' in *Oxford Dictionary of National Biography* Oxford, Oxford University Press (2004)

A.R. Favazza, *Bodies under Siege: Self-Mutilation and Body Modification in Culture and Psychiatry* 2nd ed. Baltimore, Johns Hopkins University Press (1996)

A.R. Favazza, *Bodies Under Siege: Self-Mutilation, Nonsuicidal Self-Injury, and Body Modification in Culture and Psychiatry* 3rd ed. Baltimore, Johns Hopkins University Press (2011)

J.J. Fleminger and B.L. Mallett, 'Psychiatric Referrals from Medical and Surgical Wards' *British Journal of Psychiatry* 108 (1962): 183–90

M. Foucault, *Politics, Philosophy, Culture: Interviews and Other Writings, 1977–1984* London, Routledge (1988)

M. Foucault, *The Essential Works of Michel Foucault, 1954–1984*. Vol. 2, *Aesthetics* London, Penguin (2000)

M. Foucault, *The Birth of Biopolitics: Lectures at the College de France, 1978–1979* A.I. Davidson (ed.) and Graham Burchell (trans.) Basingstoke, Palgrave Macmillan (2008)

H.L. Freeman, 'Psychiatry in the National Health Service 1948–1998' *British Journal of Psychiatry* 175 (1999): 3–11

J. Fry, *Casualty Services and Their Setting: A Study in Medical Care* Nuffield Provincial Hospitals Trust (1960)

A.D. Gardner, 'Debauchery of Honest Words' *British Medical Journal* 1, 4760 (1952): 715

A.R. Gardner and A.J. Gardner, 'Self-mutilation, Obsessionality and Narcissism' *British Journal of Psychiatry* 127 (1975): 127–32

F. Gardner, *Self-Harm: A Psychotherapeutic Approach* Hove, Brunner-Routledge (2001)

R. Gardner, R. Hanka, V.C. O'Brien, A.J. Page and R. Rees, 'Psychological and Social Evaluation in Cases of Deliberate Self-poisoning Admitted to a General Hospital' *British Medical Journal* 2, 6102 (1977): 1567–70

R. Gardner, R. Hanka, B. Evison, P.M. Mountford, V.C. O'Brien and S.J. Roberts, 'Consultation-liaison Scheme for Self-poisoned Patients in a General Hospital' *British Medical Journal* 2, 6149 (1978): 1392–4

S. Gilman, 'From Psychiatric Symptom to Diagnostic Category: Self-harm from the Victorians to DSM-5' *History of Psychiatry* 24 (2013): 148–65

T.H. Goodwin, 'The Casualty Clearing Stations' *Journal of the American Medical Association* 69(8) (1917): 636–7

M. Gorsky, 'The British National Health Service 1948–2008: A Review of the Historiography' *Social History of Medicine* 21(3) (2008): 437–60

M. Gorsky and J. Mohan, 'London's Voluntary Hospitals in the Interwar Period: Growth, Transformation, or Crisis?' *Nonprofit and voluntary sector quarterly* 30(2) (2001): 247–75

H. Graff and R. Mallin, 'The Syndrome of the Wrist Cutter' *American Journal of Psychiatry* 124 (1967): 36–42

S. Greer and J.C. Gunn, 'Attempted Suicides from Intact and Broken Parental Homes' *British Medical Journal* 2, 5526 (1966): 1355–7

S. Greer, J.C. Gunn and K.M. Koller, 'Aetiological Factors in Attempted Suicide' *British Medical Journal* 2, 5526 (1966): 1352–5

H. Grunebaum and G. Klerman, 'Wrist Slashing' *American Journal of Psychiatry* 124 (1967): 527–34

H.R. Guly, *A History of Accident and Emergency Medicine, 1948–2004* Basingstoke, Palgrave Macmillan (2005)

I. Hacking, *The Taming of Chance* Cambridge, Cambridge University Press (1990)

I. Hacking, *Rewriting the Soul* Princeton, Princeton University Press (1995)

I. Hacking, *Historical Ontology* London, Harvard University Press (2002)

I. Hacking, 'Making up People' in *Beyond the Body Proper: Reading the Anthropology of Material Life* M. Lock and J. Farquhar (eds) Durham, NC, Duke University Press (2007): 150–63

Ali Haggett, 'Housewives, Neuroses and the Domestic Environment in Britain, 1945–1970' in *Health and the Modern Home* M. Jackson (ed.) Abingdon, Routledge (2007): 84–110

M. Halewood, *Rethinking the Social through Durkheim, Marx and Whitehead* London, Anthem Press (2014)

M.B. Hall and F. Hopkins, 'Parental Loss and Child Guidance' *Archives of Disease in Childhood* 11, 64 (1936): 187–94

Hansard: Official Report of Debates in Parliament online at: http://hansard.millbanksystems.com accessed 29 January 2015

G. Harkins, 'Foucault, the Family and the Cold Monster of Neoliberalism' *Foucault the Family and Politics* R. Duschinsky and L.A. Rocha (eds) Basingstoke, Palgrave Macillan (2012): 82–117

T. Harrison, *Bion, Rickman, Foulkes and the Northfield Experiments: Advancing on a Different Front* London, Jessica Kingsley (2000)

C. Hassall and W.H. Trethowan, 'Suicide in Birmingham' *British Medical Journal* 1, 5802 (1972): 717–8

K. Hawton, 'Excessive Use of Psychiatric Services' *British Medical Journal* 3, 5983 (1975): 595–6

K. Hawton, 'Deliberate Self-poisoning and Self-injury in the Psychiatric Hospital' *British Journal of Medical Psychology* 51(3) (1978): 253–9

K. Hawton and J. Catalán, *Attempted Suicide* 1st ed. Oxford, Oxford University Press (1982)

K. Hawton and J. Catalán, *Attempted Suicide* 2nd ed. Oxford, Oxford University Press (1987)

K. Hawton and M. Goldacre, 'Hospital Admissions for Adverse Effects of Medicinal Agents (Mainly Self-Poisoning) Among Adolescents in the Oxford Region' *British Journal of Psychiatry* 141 (1982): 166–70

K. Hawton, J. O'Grady, M. Osborn and D. Cole 'Adolescents who Take Overdoses: Their Characteristics, Problems and Contacts with Helping Agencies' *British Journal of Psychiatry* 140 (1982): 118–23

K. Hawton, K. Rodham, E. Evans and R. Weatherall, 'Deliberate Self Harm in Adolescents: Self Report Survey in Schools in England' *British Medical Journal* 325 (2002): 1207–11.

K. Hawton, K.E.A. Saunders and R.C. O'Connor, 'Self-harm and Suicide in Adolescents' *Lancet* 379, 9834 (2012): 2373–82

E. Heimler, *Mental Illness and Social Work* Harmondsworth, Penguin (1967)

D.K. Henderson and R.D. Gillespie, *A Text-Book of Psychiatry for Students and Practitioners* 1st ed. London, Humphrey Milford (1927)

D.K. Henderson and R.D. Gillespie, *A Text-Book of Psychiatry for Students and Practitioners* 6th ed. Oxford, Oxford University Press (1944)

D.K. Henderson and I.R.C. Batchelor, *Henderson and Gillespie's Textbook of Psychiatry* 9th ed. London, Oxford University Press (1962)

D.K. Henderson, *The Evolution of Psychiatry in Scotland* Edinburgh, E. & S. Livingstone (1964)

P. Hennessy, *Having It So Good: Britain in the Fifties* London, Allen Lane (2006)

T.A. Holding, D. Buglass, J.C. Duffy and N. Kreitman, 'Parasuicide in Edinburgh – A Seven-Year Review 1968–1974' *British Journal of Psychiatry* 130 (1977): 534–43

T.H. Holmes and R.H. Rahe, 'The Social Readjustment Rating Scale' *Journal of Psychosomatic Research* 11(2) (1967): 213–18

F. Hopkins, 'Attempted Suicide: An Investigation' *Journal of Mental Science* 83 (1937): 71–94

F. Hopkins, 'Admissions to Mental Observation Wards During War' *British Medical Journal* 1, 4289 (1943): 358

N.K. Hunnybun, 'Psychiatric Social Work' in *Social Case-Work in Great Britain* C. Morris (ed.) London: Faber and Faber (1950)

J. Hyman, *Women Living With Self-Injury* Philadelphia, Temple University Press (1999)

E. Irvine, 'Psychiatric Social Work' in E. Younghusband, *Social Work in Britain, 1950–1975: A Follow-Up Study* London, Allen & Unwin (1978)

R. Jack, *Women and Attempted Suicide* London, Lawrence Erlbaum (1992)

M. Jackson, *The Age of Stress: Science and the Search for Stability* Cambridge, Cambridge University Press (2013)

C. Jacobson and K. Batejan, 'Comprehensive Theoretical Models of Nonsuicidal Self-Injury' in *Oxford Handbook of Suicide and Self-Injury* M. Nock (ed.) Oxford, Oxford University Press (2014)

A. Strauss, L. Schatzman, R. Bucher, D. Ehrlich and M. Sabshin, *Psychiatric Ideologies and Institutions* London, Collier-Macmillan Limited (1964)

Å. Jansson, 'From Statistics to Diagnostics: Medical Certificates, Melancholia, and "Suicidal Propensities" in Victorian Medicine' *Journal of Social History* 46(3) (2013): 716–31

M. Jarvis, *Conservative Governments, Morality and Social Change in Affluent Britain, 1957–64* Manchester, Manchester University Press (2005)

F.A. Jenner, 'Stengel, Erwin (1902–1973)' in *Oxford Dictionary of National Biography* Oxford, Oxford University Press (2004)

Joint Committee of the British Medical Association and the Magistrates' Association, 'The Law and Practice in Relation to Attempted Suicide in England and Wales' London, British Medical Association and Magistrates' Association (1958)

M. Jones, *Social Psychiatry in Practice: The Idea of the Therapeutic Community* Harmondsworth, Penguin (1968)

E. Jones, S. Rahman and R. Woolven, 'The Maudsley Hospital: Design and Strategic Direction, 1923–1939' *Medical History* 51(3) (2007): 357–78

J.S. Kafka, 'The Body as Transitional Object: A Psychoanalytic Study of a Self-mutilating Patient' *British Journal of Medical Psychology* 42 (1969): 207–12

D. Keay, interviewing Margaret Thatcher, 'Aids, Education and the Year 2000!' *Woman's Own* (31 October 1987) Transcript online at: http://www.margaret-thatcher.org/document/106689 accessed 28 January 2015

J.G. Keilp, M.A. Oquendo, B.H. Stanley, A.K. Burke, T.B. Cooper, K.M. Malone and J.J. Mann, 'Future Suicide Attempt and Responses to Serotonergic Challenge.' *Neuropsychopharmacology* 35 (2010): 1063–72

G.A. Kelly, *The Psychology of Personal Constructs* 2 vols, New York, Norton (1955)

N. Kessel, 'Psychiatric Morbidity in a London General Practice' *British Journal of Preventive & Social Medicine* 14 (1960): 16–22

N. Kessel, 'Neurosis & the N.H.S.' *The Twentieth Century* CLXXII, 1015 (1962): 55–65

N. Kessel, 'Attempted Suicide' *Medical World* 97 (1962): 312–22

N. Kessel, 'Attempted Suicide' *Lancet* 281, 7278 (1963): 448

N. Kessel, 'Self-Poisoning – Part I' *British Medical Journal* 2, 5473 (1965): 1265–70

N. Kessel, 'Self-Poisoning – Part II' *British Medical Journal* 2, 5474 (1965): 1336–40

N. Kessel, 'The Respectability of Self-Poisoning and the Fashion of Survival' *Journal of Psychosomatic Research* 10(1) (1966): 29–36

N. Kessel, W. McCulloch, J. Hendry, D. Leslie, I. Wallace and R. Webster, 'Hospital Management of Attempted Suicide in Edinburgh' *Scottish Medical Journal* 9 (1964): 333–4

N. Kessel, W. McCulloch and E. Simpson, 'Psychiatric Service in a Centre for the Treatment of Poisoning' *British Medical Journal* 2, 5363 (1963): 985–8

N. Kessel and G. Grossman, 'Suicide in Alcoholics' *British Medical Journal* 2, 5268 (1961): 1671–2

N. Kessel and E.M. Lee, 'Attempted Suicide in Edinburgh' *Scottish Medical Journal* 7 (1962): 130–5

N. Kessel and W. McCulloch, 'Repeated Acts of Self-Poisoning and Self-Injury' *Proceedings of the Royal Society of Medicine* 59(2) (1966): 89–92

N. Kessel and M. Shepherd, 'Neurosis in Hospital and General Practice' *Journal of Mental Science* 108 (1962): 159–66

N. Kessel and H.J. Walton, *Alcoholism* Harmondsworth, Penguin (1965)

A. Kraus, G. Valerius, E. Seifritz, M. Ruf, J.D. Bremner, M. Bohus and C. Schmahl, 'Script-Driven Imagery of Self-injurious Behavior in Patients with Borderline Personality Disorder: A Pilot FMRI Study' *Acta Psychiatrica Scandinavica* 121(1) (2010): 41–51

I.S. Kreeger, 'Initial Assessment of Suicidal Risk' *Proceedings of the Royal Society of Medicine* 59(2) (1966): 92–6

N. Kreitman, 'Mental Disorder in Married Couples' *British Journal of Psychiatry* 108 (1962): 438–46

N. Kreitman, 'The Patient's Spouse' *British Journal of Psychiatry* 110 (1964): 159–73

N. Kreitman, A.E. Philip, S. Greer and C.R. Bagley, 'Parasuicide' *British Journal of Psychiatry* 115 (1969): 746–7

N. Kreitman, 'The Coal Gas Story. United Kingdom Suicide Rates, 1960–71' *British Journal of Preventive & Social Medicine* 30(2) (1976): 86–93

N. Kreitman, (ed.) *Parasuicide* London, John Wiley & Sons (1977)

N. Kreitman, 'Can Suicide and Parasuicide Be Prevented?' *Journal of the Royal Society of Medicine* 82(11) (1989): 648–52

R. Krohn, 'Why Are Graphs So Central in Science?' *Biology and Philosophy* 6(2) (1991): 181–203

H.I. Kushner, 'Women and Suicide in Historical Perspective' *Signs* 10(3) (1985): 537–52

H.I. Kushner, 'Biochemistry, Suicide and History: Possibilities and Problems' *Journal of Interdisciplinary History* 16(1) (1985): 69–85

B. Latour and S. Woolgar, *Laboratory Life: The Social Construction of Scientific Facts* London, Sage Publications (1979)

A. A.H. Lawson and I. Mitchell, 'Patients with acute poisoning seen in a general medical unit (1960–71)' *British Medical Journal* 4, 5833 (1972): 153–6

J.E. Lennard-Jones and R.A.J. Asher, 'Why Do They Do It? A Study of Pseudocide' *Lancet* 1, 7083 (1959): 1138–40

H. Lester and J. Glasby, *Mental Health Policy and Practice* Basingstoke, Palgrave Macmillan (2006)

I. Leveson, 'Evolution of Psychiatry in a Clinician's Lifetime' *Transactions and Report/Liverpool Medical Institution* (1968): 23–5

A. Lewis, 'Social Aspects of Psychiatry: Part I' *Edinburgh Medical Journal* 58, 5 (1951): 214–30

J. Lewis, 'Public Institution and Private Relationship: Marriage and Marriage Guidance, 1920–1968' *Twentieth Century British History* 1, 3 (1990): 233–63

J.B.S. Lewis, 'Mental Health Services' *British Medical Journal* 2, 4900 (1954): 1354–5

R. Leys, 'How Did Fear Become a Scientific Object and What Kind of Object Is It?' *Representations* 110 (2010): 66–104

J.C. Little, 'A Rational Plan for Integration of Psychiatric Services to an Urban Community' *Lancet* 2, 7318 (1963): 1159–60

M. Littlefield and J. Johnson (eds) *The Neuroscientific Turn: Transdisciplinarity in the Age of the Brain* Ann Arbor, University of Michigan Press (2012)

G.G. Lloyd and R.A. Mayou, 'Liaison Psychiatry or Psychological Medicine?' *British Journal of Psychiatry* 183 (2003): 5–7

S. Locket, 'Barbiturate Deaths' *British Medical Journal* 2, 5162 (1959): 1332

V. Long, '"Often There Is a Good Deal to Be Done, but Socially Rather Than Medically": The Psychiatric Social Worker as Social Therapist, 1945–70' *Medical History* 55(2) (2011): 223–39

A. Lovell and E. Susser (eds) *International Journal of Epidemiology* Special Issue on the History of Psychiatric Epidemiology (2014)

T.G. Lowden, *The Casualty Department* London, E. & S. Livingstone (1955)

T.G. Lowden, 'The Casualty Department. I. The Work and the Staff' *Lancet* 270, 6929 (1956): 955–6

J. M. Mackintosh, *The War and Mental Health in England* New York, Commonwealth Fund (1944)

W.S. Maclay, 'After the Mental Health Act: An Appraisal of English Psychiatry' *Mental Hospitals* 14 (1963): 100–6

W.S. Maclay, 'The Adolf Meyer Lecture: A Mental Health Service' *The American Journal of Psychiatry* 120 (1963): 209–17

W.S. Maclay, 'Trends in the British Mental Health Service' in *Trends in the Mental Health Services: A Symposium of Original and Reprinted Papers* H.L. Freeman and W.A.J. Farndale (eds) Oxford, Pergamon Press (1963): 3–11

J.G. Macleod and H.J. Walton, 'Liaison between Physicians and Psychiatrists in a Teaching Hospital' *Lancet* 294, 7624 (1969): 789–92

T.F. Main, 'The Ailment' *Medical Psychology* 30(3) (1957): 129–45

J. Marshall, 'Mental Health Services' *British Medical Journal* 2, 4902 (1954): 1484–5

H. Matthew, 'Poisoning in the Home by Medicaments' *British Medical Journal* 2, 5517 (1966): 788–90

H. Matthew, A.T. Proudfoot, S.S. Brown and R.C. Aitken, 'Acute Poisoning: Organization and Work-Load of a Treatment Centre' *British Medical Journal* 3, 5669 (1969): 489–92

H.J.S. Matthew and A.A.H. Lawson, *Treatment of Common Acute Poisonings* Edinburgh, E. & S. Livingstone (1967)

P.C. Matthews 'Epidemic Self-Injury in an Adolescent Unit' *International Journal of Social Psychiatry* 14 (1968): 125–33

W. Mayer-Gross, M. Roth and E.T.O. Slater, *Clinical Psychiatry* 1st ed. London, Cassell & Co. (1954)

W. Mayer-Gross, M. Roth and E.T.O. Slater, *Clinical Psychiatry* 2nd ed. London, Cassell & Co. (1960)

R. Mayou, 'The History of General Hospital Psychiatry' *British Journal of Psychiatry* 155 (1989): 764–76

P. McCormick, *Cut* London, Collins Flamingo (2009)

W. McCulloch and A.E. Philip, *Suicidal Behaviour* Oxford, Pergamon (1972)

C.P. McEvedy and A.W. Beard, 'Royal Free Epidemic of 1955: A Reconsideration' *British Medical Journal* 1, 5687 (1970): 7–11

C.P. McEvedy, A. Griffith and T. Hall, 'Two School Epidemics' *British Medical Journal* 2, 5525 (1966): 1300–2

D.W. McKerracher, T. Loughnane and R.A. Watson, 'Self-Mutilation in Female Psychopaths' *British Journal of Psychiatry* 114 (1968): 829–32

D.W. McKerracher, D.R.K. Street and L.J. Segal, 'A Comparison of the Behaviour Problems Presented by Male and Female Subnormal Offenders' *British Journal of Psychiatry* 112 (1966): 891–97

L. McKinstry, 'This "Epidemic" Is All Selfishness' *Daily Telegraph* (01 August 2004)

K. McMillan, 'Under a Redescription' *History of the Human Sciences* 16(2) (2003): 129–50

T. McShane, *Blades, Blood and Bandages* Basingstoke, Palgrave Macmillan (2012)

A. McSmith, *No Such Thing as Society* London, Constable (2011)

A. McSmith, 'Mental Health Act, 1959' London: HMSO (1959)

G.D. Middleton, D.W. Ashby and F. Clark, 'An Analysis of Attempted Suicide in an Urban Industrial District' *Practitioner* 187 (1961): 776–82

W.M. Millar, G. Innes and G.A. Sharp, 'Hospital and Outpatient Clinics: The Design of a Reporting System and the Difficulties to Be Expected in the Execution' in *The Burden on the Community: The Epidemiology of Mental Illness a Symposium* (1962)

C. Millard, 'Reinventing Intention: "Self-Harm" and the "Cry for Help" in Post-War Britain' *Current Opinion in Psychiatry* 25(6) (2012): 503–7

C. Millard, 'Making the Cut: The Production of "Self-harm" in Anglo-Saxon Psychiatry' *History of the Human Sciences* 26(2) (2013): 126–50

C. Millard, 'Stress Attempted Suicide and The Social' in *Stress in Post-War Britain* M. Jackson (ed.) Studies for the Society for the Social History of Medicine, Pickering and Chatto (forthcoming 2015)

C. Millard and S. Wessely, 'Parity of Esteem Between Physical and Mental Health' *British Medical Journal* 349, 7894 (2014): 6821

Ministry of Health, 'Accident and Emergency Services: Report of a Sub-Committee (Platt Report)' London, HMSO (1962)

Ministry of Health, 'A Hospital Plan for England and Wales' London, HMSO (1962)

Ministry of Health, 'Report of the Working Party on Social Workers in the Local Authority Health and Welfare Services (Younghusband Report)' London, HMSO (1959)

Moorhead, J. 'Self-harm among Children is on the Rise, But it's Not Just the Victims Who We Need to Support' *Independent* (11 August 2014) online at: http://www.independent.co.uk/voices/comment/selfharm-among-children-is-on-the-rise-but-its-not-just-the-victims-who-we-need-to-support-9662252.html accessed 12 August 2014

H.G. Morgan, H. Pocock and S. Pottle, 'The Urban Distribution of Non-Fatal Deliberate Self-Harm' *British Journal of Psychiatry* 126 (1975): 319–28

H.G. Morgan, C.J. Burns-Cox, H. Pocock and S. Pottle, 'Deliberate Self-Harm: Clinical and Socio-Economic Characteristics of 368 Patients' *British Journal of Psychiatry* 127 (1975): 564–74

H.G. Morgan, *Death Wishes? The Understanding and Management of Deliberate Self-Harm* Chichester, John Wiley & Sons (1979)

P.D. Moss and C.P. McEvedy, 'An Epidemic of Overbreathing Among Schoolgirls' *British Medical Journal* 2, 5525 (1966): 1295–300

A. Munro and J.W. McCulloch, *Psychiatry for Social Workers* 2nd ed. Oxford, Pergamon Press (1975)

S. Nelson and H. Grunebaum, 'A Follow-up Study of Wrist-slashers' *American Journal of Psychiatry* 127 (1971): 1345–9

W.L. Neustatter, *Psychological Disorder and Crime* London, Christopher Johnson (1953)

J.G.B. Newson-Smith, 'Use of Social Workers as Alternatives to Psychiatrists in Assessing Parasuicide' in *The Suicide Syndrome* R. Farmer and S. Hirsch (eds) London, Croom Helm (1980): 215–25

V. Norris, *Mental Illness in London* London, Chapman & Hall (1959)

D. Odlum, P. Epps, I.R.C. Batchelor and I.M.H. McAdam, 'Discussion on the Legal Aspects of Suicidal Acts' *Proceedings of the Royal Society of Medicine* 51(4) (1958): 297–303

D. Offer and P. Barglow, 'Adolescent and Young Adult Self-Mutilation Incidents in a General Psychiatric Hospital' *Archives of General Psychiatry* 3 (1960): 194–204

D. Pallis, A. Langley and D. Birtchnell, 'Excessive Use of Psychiatric Services by Suicidal Patients' *British Medical Journal* 3, 5977 (1975): 216–18

P.-N. Pao, 'The Syndrome of Delicate Self-cutting' *British Journal of Medical Psychology* 42 (1969): 195–205

M. Parascandola, 'Epidemiology in Transition: Tobacco and Lung Cancer in the 1950s' in *Body Counts: Medical Quantification in Historical and Sociological Perspective*, G. Jorland, A. Opinel and G. Weisz (eds) Ithaca, McGill-Queens University Press (2005)

D. Parkin and E. Stengel, 'Incidence of Suicidal Attempts in an Urban Community' *British Medical Journal* 2, 5454 (1965): 133–8

E.S. Paykel, B.A. Prusoff and J.K. Myers, 'Suicide Attempts and Recent Life Events. A Controlled Comparison' *Archives of General Psychiatry* 32(3) (1975): 327–33

N. Pengelly, B. Ford, P. Blenkiron and S. Reilly, 'Harm Minimisation after Repeated Self-harm: Development of a Trust Handbook' *Psychiatric Bulletin* 32 (2008): 60–3

L.S. Penrose, 'Mental Illness in Husband and Wife: A Contribution to the Study of Assortive Mating in Man' *Psychiatric Quarterly* 18 (1944): 161–6

E.U.H. Pentreath and E.C. Dax, 'Mental Observation Wards: A Discussion of Their Work and Its Objects' *Journal of Mental Science* 83 (1937): 347–65

Baron Percy of Newcastle, 'The Report of the Royal Commission of the Law Relating to Mental Illness and Mental Deficiency' London, HMSO (1957)

U.H. Peters, 'The Emigration of German Psychiatrists to Britain' in *150 Years of Psychiatry: The Aftermath* H.L. Freeman and G.E. Berrios (eds) London, Athlone (1996): 565–80

M. Pickersgill, 'Between Soma and Society: Neuroscience and the Ontology of Psychopathy' *BioSocieties* 4 (2009): 45–60

M. Pickersgill, 'Psyche, Soma, and Science Studies: New Directions in the Sociology of Mental Health and Illness' *Journal of Mental Health* 19(4) (2010): 382–92

M. Pickersgill, 'Social Life of the Brain' *Current Sociology* 61(3) (2013): 322–40

J. Pickstone, 'Psychiatry in District General Hospitals' in *Medical Innovations in Historical Perspective* J. Pickstone (ed.) (1992): 185–99

R. Porter, *The Greatest Benefit to Mankind: A Medical History of Humanity from Antiquity to the Present* London, Fontana Press (1999)

R. Porter, 'The Patient's View: Doing Medical History from Below' *Theory and Society* 14(2) (1985): 175–98

F. Post, 'Mental Breakdown in Old Age' *British Medical Journal* 1, 4704 (1951): 436–40

F. Post and J. Wardle, 'Family Neurosis and Family Psychosis' *British Journal of Psychiatry* 108 (1962): 147–58

E. Powell, *Emerging Patterns for the Mental Health Services and the Public* London, National Association for Mental Health (1961)

A. Proudfoot and L.F. Lescott, 'Henry Matthew: The Father of Modern Clinical Toxicology' *Journal of the Royal College of Physicians of Edinburgh* 39 (2009): 357–61

C. Rainfield, *Scars* Lodi, New Jersey, Westside Books (2011)

K.S. Rajan, *Biocapital: The Constitution of Postgenomic Life* Durham, NC, Duke University Press (2006)

E. Ramsden and D. Wilson, 'The Nature of Suicide: Science and the Self-Destructive Animal' *Endeavour* 34(1) (2010): 21–4

J.R. Rees, *The Shaping of Psychiatry by War* London, Chapman & Hall (1945)

D.D. Reid, *Epidemiological Methods in the Study of Mental Disorders* Geneva: World Health Organization (1960)

C. Richmond, 'Colin McEvedy' *British Medical Journal* 331 (2005): 847

C. Rinzler and D. Shapiro, 'Wrist-Cutting and Suicide' *Journal of Mount Sinai Hospital, New York* 35 (1968): 485–8

G. Rivett, 'The Voluntary Hospitals' in *Development of the London Hospital System 1823–1992* online at: http://www.nhshistory.net/voluntary_hospitals.htm accessed 30 January 2015

J. Roberts and D. Hooper, 'The Natural History of Attempted Suicide in Bristol' *British Journal of Medical Psychology* 42(4) (1969): 303–12

K. Robinson, 'The Public and Mental Health' in *Trends in the Mental Health Services* H.L. Freeman and W.A.J. Farndale (eds) Oxford, Pergamon (1963): 12–18

A. Rogers and D. Pilgrim, *Mental Health Policy in Britain* 2nd ed. London, Macmillan (2001)

H.R. Rollin, 'Social and Legal Repercussions of the Mental Health Act, 1959' *British Medical Journal* 1, 5333 (1963): 786–8

H.R. Rollin, 'Carstairs, George Morrisson (1916–1991)' in *Oxford Dictionary of National Biography* London, Oxford University Press (2004)

N. Rose, 'The Death of the Social? Re-figuring the Territory of Government' *Economy and Society* 25(3) (1996): 327–56

N. Rose, *Governing the Soul: The Shaping of the Private Self* London, Free Association Press (1999)

N. Rose, *Politics of Life Itself: Biomedicine, Power and Subjectivity in the Twenty-First Century* Princeton, NJ, Princeton University Press (2007)

H. Rose and S. Rose, *Genes, Cells and Brains: The Promethean Promises of the New Biology* London, Verso (2012)

R. Rosenthal, C. Rinzler, R. Wallsch and E. Klausner, 'Wrist-cutting Syndrome: The Meaning of a Gesture' *American Journal of Psychiatry* 128 (1972): 1363–8

A. Roy, 'Self-Mutilation' *British Journal of Medical Psychology* 51 (1978): 201–3.

Royal College of Psychiatrists, 'Self-harm, Suicide and Risk: Helping People Who Self-harm' College Report CR158 (2010) online at: http://www.rcpsych.ac.uk/files/pdfversion/cr158.pdf accessed 30 January 2015

J. Ruesch and G. Bateson, *Communication: The Social Matrix of Psychiatry* New York, W.W. Norton (1951)

P. Sainsbury and J. Grad, 'Evaluation of Treatment and Services' in 'The Burden on the Community: The Epidemiology of Mental Illness: A Symposium' (1962)

Samaritans, 'Samaritans History' online at: http://www.samaritans.org/about_samaritans/governance_and_history/samaritans_history.aspx accessed 11 July 2012.

J. Sandler, 'Studies in Psychopathology using a Self-Assessment Inventory. I. The Development and Construction of the Inventory' *British Journal of Medical Psychology* 27 (1954): 147–52

S.R. Schroeder, M.L. Oster-Granite and T. Thompson, *Self-Injurious Behavior: Gene-Brain-Behavior-Relationships* American Psychological Association, Washington, D.C. (2002)

A.B. Sclare and C.M. Hamilton, 'Attempted Suicide in Glasgow' *British Journal of Psychiatry* 109 (1963): 609–15

J.W. Scott, 'The Evidence of Experience' *Critical Enquiry* 17(4) (1991): 773–97

J.W. Scott, *Gender and the Politics of History* rev. ed. New York, Columbia University Press (1999)

J.W. Scott, 'History-writing as Critique' in K. Jenkins, S. Morgan and A. Munslow (eds), *Manifestos for History* Abingdon: Routledge (2007)

A.T. Scull, *Decarceration: Community Treatment and the Deviant: A Radical View* London, Prentice-Hall (1977)

C.P. Seager, *Psychiatry for Nurses Social Workers and Occupational Therapists* London, Heinemann Medical (1968)

P. Sedgwick, *Psychopolitics* London, Pluto Press (1982)

H. Selye, *Stress of Life* London, McGraw-Hill (1956)

S. Shapin and S. Schaffer, *Leviathan and the Air-Pump: Hobbes, Boyle and the Experimental Life* Princeton, NJ, Princeton University Press (1985)

B. Shephard, *A War of Nerves: Soldiers and Psychiatrists in the Twentieth Century* Cambridge, MA, Harvard University Press (2001)

A. Shepherd and D. Wright, 'Madness, Suicide and the Victorian Asylum: Attempted Self-Murder in the Age of Non-Restraint' *Medical History* 46(2) (2002): 175–96

M. Shepherd, 'Morbid Jealousy: Some Clinical and Social Aspects of a Psychiatric Symptom' *The British Journal of Psychiatry* 107 (1961): 687–753

M. Shepherd, 'Epidemiology and Clinical Psychiatry' *British Journal of Psychiatry* 133 (1978): 289–98

M. Shepherd, 'The Impact of Germanic Refugees on Twentieth-Century British Psychiatry' *Social History of Medicine* 22(3) (2009): 461–9

M. Shepherd, A.C. Brown, B. Cooper and G. Kalton, *Psychiatric Illness in General Practice* London, Oxford University Press (1966)

E. Showalter, *The Female Malady: Women, Madness and English Culture 1830–1980* London, Virago (1987)

M. Sim, 'Psychological Aspects of Poisoning' in G. Cumming, *The Medical Management of Acute Poisoning* London, Cassell (1961)

M. Sim, *Guide to Psychiatry* 1st ed. Edinburgh and London, E & S Livingstone (1963)

M. Sim, *Guide to Psychiatry* 2nd ed. Edinburgh and London, E & S Livingstone (1968)

M. Sim, *Guide to Psychiatry* 3rd ed. (With a chapter on legal aspects of psychiatry in the United States of America by John Donnelly) Edinburgh and London, Churchill Livingstone (1974)

M. Sim, *Guide to Psychiatry* 4th ed. (With a chapter on legal aspects of psychiatry in the United States of America by John Donnelly) Edinburgh & London, Churchill Livingstone (1981)

M.A. Simpson, *Medical Education: A Critical Approach* Butterworths, London (1972)

M.A. Simpson, 'Phenomenology of Self-Mutilation in a General Hospital Setting' *Canadian Psychiatric Association Journal* 20(6) (1975): 429–34

M.A. Simpson, 'Self Mutilation and Suicide' in *Suicidology: Contemporary Developments* E.S. Shneidman (ed.) New York, Grune and Stratton (1976)

M.A. Simpson 'Self-Mutilation and Borderline Syndrome' *Dynamische Psychiatrie* 1 (1977): 42–8

K. Skegg, 'Self-harm' *Lancet* 366, 9495 (2005): 1471–83

I. Skottowe, 'Discussion: Observation Units' *Proceedings of the Royal Society of Medicine* 33 (1940): 732–4

E.T.O. Slater and M. Woodside, *Patterns of Marriage: A Study of Marriage Relationships in the Urban Working Classes* London, Cassell (1951)

E.T.O. Slater and M. Roth, *Clinical Psychiatry* 3rd ed. London, Baillière, Tindall & Cassell (1969)

E.T.O. Slater, 'Psychiatry in the 'Thirties' *Contemporary Review* 226, 1309 (1975): 70–5

D. Smail, *On Deep History and the Brain* Berkeley: University of California Press (2008)

A.J. Smith, 'Self-Poisoning with Drugs: A Worsening Situation' *British Medical Journal* 4, 5833 (1972): 157–9

R.P. Snaith and S. Jacobson, 'The Observation Ward and the Psychiatric Emergency' *British Journal of Psychiatry* 111 (1965): 18–26

D. Stafford-Clark, 'Attempted Suicide' *Lancet* 281, 7278 (1963): 448–9

Standing Medical Advisory Committee of the Central Health Services Council, 'Report of the Sub Committee: Emergency Treatment in Hospital of Cases of Acute Poisoning' [Atkins Report] London, HMSO (1962)

Standing Medical Advisory Committee of the Central Health Services Council, 'Hospital Treatment of Acute Poisoning. Report of the Joint Sub-Committee of the Standing Medical Advisory Committees' [Hill Report] London, HMSO (1968)

B. Stanley and L. Siever, 'The Interpersonal Dimension of Borderline Personality Disorder: Toward a Neuropeptide Model' *American Journal of Psychiatry* 167 (2010): 24–39

A.H. Stanton and M.S. Schwartz, *The Mental Hospital* London, Tavistock Publications (1954)

M.E. Staub, *Madness Is Civilization: When the Diagnosis Was Social, 1948–1980* London, University of Chicago Press (2011)

E. Stengel, 'On the Aetiology of the Fugue States' *Journal of Mental Science* 87 (1941): 572–99

E. Stengel, 'Further Studies on Pathological Wandering (Fugues with the Impulse to Wander)' *Journal of Mental Science* 89 (1943): 224–41

E. Stengel, 'Suicide' in *Recent Progress in Psychiatry* G.W.T.H. Fleming, A. Walk and P.K. McCowan (eds) London, J. & A. Churchill Ltd (1950): 691–703

E. Stengel, 'Enquiries into Attempted Suicide (Abridged)' *Proceedings of the Royal Society of Medicine* 45 (1952): 613–20

E. Stengel, 'Attempted Suicide and the Law' *Medico-Legal Journal* 27 (1959): 114–20

E. Stengel, 'The National Health Service and the Suicide Problem' in *Sociological Review Monograph No.5: Sociology and Medicine: Studies within the Framework of the British National Health Service* P. Halmos (ed.) Keele: University of Keele (1962): 195–205

E. Stengel, 'Attempted Suicide: Its Management in the General Hospital' *Lancet* 1, 7275 (1963): 233–5

E. Stengel, *Suicide and Attempted Suicide* Harmondsworth, Penguin Books (1964)

E. Stengel, N.G. Cook and I.S. Kreeger, *Attempted Suicide: Its Social Significance and Effects* London, Chapman and Hall (1958)

J. Stewart, '"I Thought You Would Want to Come and See His Home": Child Guidance and Psychiatric Social Work in Inter-War Britain' in *Health and the Modern Home* M. Jackson (ed.) London, Routledge (2007): 111–27

A. Storr, 'Bowlby, (Edward) John Mostyn (1907–1990)' in *Oxford Dictionary of National Biography* Oxford, Oxford University Press (2004)

J. Sutton, *Healing the Hurt Within: Understand Self-Injury and Self-Harm, and Heal the Emotional Wounds* 3rd ed. Oxford, How To Books (2007)

T.S. Szasz, *The Myth of Mental Illness: Foundations of a Theory of Personal Conduct* London, Paladin ([1961] 1972)

D. Tait, 'Norman Kreitman in Conversation with David Tait' *Psychiatric Bulletin* 19 (1995): 296–301

D. Tantam and N. Huband, *Understanding Repeated Self-Injury: A Multidisciplinary Approach* Basingstoke, Palgrave Macmillan (2009)

M. Thomson, *Psychological Subjects: Identity, Culture, and Health in Twentieth-Century Britain* Oxford, Oxford University Press (2006)

N. Timms, *Psychiatric Social Work in Great Britain, 1939–1962* London, Routledge & Kegan Paul (1964)

R.M. Titmuss, 'Community Care: Fact or Fiction?' in *Commitment to Welfare* R.M. Titmuss (ed.) London: George Allen and Unwin ([1961] 1968): 221–5

J. Toms, *Mental Hygiene and Psychiatry in Modern Britain* Basingstoke, Palgrave Macmillan (2013)

G.C. Tooth and E.M. Brooke, 'Trends in the Mental Hospital Population and Their Effect on Future Planning' *Lancet* 1, 7179 (1961)

W.H. Trethowan, 'Suicide and Attempted Suicide' *British Medical Journal* 2, 6185 (1979): 319–20

R. Turner and H. Morgan, 'Patterns of Health Care in Non-fatal Deliberate Self-harm' *Psychological Medicine* 9(3) (1979): 487–92

M. Vicedo, 'The Social Nature of the Mother's Tie to Her Child: John Bowlby's Theory of Attachment in Post-War America' *The British Journal for the History of Science* 44, 3 (2011): 401–26

B.M. Wagner, *Suicidal Behavior in Children and Adolescents* London, Yale University Press (2009)

J.H. Wallis, *Marriage Guidance: A New Introduction* Routledge & Keegan Paul (1968)

B. Walsh and P. Rosen, *Self-mutilation: Theory, Research and Treatment* Guildford, Guildford Press (1988)

J.P. Watson, 'Relationship between a Self-mutilating Patient and her Doctor' *Psychotherapy and Psychosomatics* 18 (1970): 67–73

C.A.H. Watts, *Depressive Disorders in the Community* Bristol, John Wright & Sons (1966)

J.C. Weaver, *A Sadly Troubled History: The Meanings of Suicide in the Modern Age* Ithaca, McGill-Queens University Press (2009)

C. Webster, 'Psychiatry and the Early National Health Service: The Role of the Mental Health Standing Advisory Committee' in *150 Years of British Psychiatry, 1841–1991* H. Freeman and G. E. Berrios (eds) London, Gaskell (1991): 103–16

P. Weindling, 'Alien Psychiatrists: The British Assimilation of Psychiatric Refugees' in *International Relations in Psychiatry: Britain, Germany, and the United States to World War II* V. Roelcke, P. Weindling and L. Westwood (eds) Rochester, University of Rochester Press (2010): 218–36

J. Welshman, 'Rhetoric and Reality: Community Care in England and Wales, 1948–74' in *Outside the Walls of the Asylum: The History of Care in the Community 1750–2000* P. Bartlett and D. Wright (eds) London: Athlone Press (1999): 204–26

G. Wilkinson, *Talking About Psychiatry* London, Gaskell (1993)

G.L. Williams, *The Sanctity of Life and the Criminal Law: On Contraception, Sterilization, Artificial Insemination, Abortion, Suicide and Euthanasia* London, Faber & Faber (1958)

A. Wilson, 'On the History of Disease Concepts: The Case of Pleurisy' *History of Science* 38 (2000): 271–319

J.K. Wing, 'Survey Methods and the Psychiatrist' in *Methods of Psychiatric Research* P. Sainsbury and N. Kreitman (eds) London, Oxford University Press (1963)

J.K. Wing, *Reasoning About Madness* Oxford, Oxford University Press (1978)

Baron Wolfenden of Westcott, *Report of the Committee on Homosexual Offences and Prostitution* London, HMSO (1957)

M. Woodside, 'Attempted Suicides Arriving at a General Hospital' *British Medical Journal* 2, 5093 (1958): 411–14

M. Woodside, 'Are Observation Wards Obsolete? A Review of One Year's Experience in an Acute Male Psychiatric Admission Unit' *British Journal of Psychiatry* 114 (1968): 1013–18

B. Wootton, *Social Science and Social Pathology* London, George Allen & Unwin (1959)

A. Young, *The Harmony of Illusions: Inventing Post-Traumatic Stress Disorder* Princeton, NJ, Princeton University Press (1995)

E.L. Younghusband, *Social Work in Britain: A Supplementary Report on the Employment and Training of Social Workers* Dunfermline, Carnegie United Kingdom Trust (1951)

E.L. Younghusband, *Social Work in Britain, 1950–1975: A Follow-up Study* London, Allen and Unwin (1978)

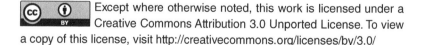 Except where otherwise noted, this work is licensed under a Creative Commons Attribution 3.0 Unported License. To view a copy of this license, visit http://creativecommons.org/licenses/by/3.0/

Index

Except where otherwise noted, this work is licensed under a
Creative Commons Attribution 3.0 Unported License. To view
a copy of this license, visit http://creativecommons.org/licenses/by/3.0/

Lightning Source UK Ltd.
Milton Keynes UK
UKOW06f0627240816

281291UK00027B/624/P